Artificial Intelligence and Human Evolution

Contextualizing AI in Human History

Ameet Joshi

Apress®

Artificial Intelligence and Human Evolution: Contextualizing AI in Human History

Ameet Joshi
REDMOND, WA, USA

ISBN-13 (pbk): 978-1-4842-9806-0   ISBN-13 (electronic): 978-1-4842-9807-7
https://doi.org/10.1007/978-1-4842-9807-7

Managing Director, Apress Media LLC: Welmoed Spahr
Acquisitions Editor: Shivangi Ramachandran
Development Editor: James Markham
Project Manager: Jessica Vakili

Distributed to the book trade worldwide by Springer Science+Business Media New York, 1 NY PLaza, New York, NY 10004. Phone 1-800-SPRINGER, fax (201) 348-4505, e-mail orders-ny@ springer-sbm.com, or visit www.springeronline.com. Apress Media, LLC is a California LLC and the sole member (owner) is Springer Science + Business Media Finance Inc (SSBM Finance Inc). SSBM Finance Inc is a **Delaware** corporation.

For information on translations, please e-mail booktranslations@springernature.com; for reprint, paperback, or audio rights, please e-mail bookpermissions@springernature.com.

Apress titles may be purchased in bulk for academic, corporate, or promotional use. eBook versions and licenses are also available for most titles. For more information, reference our Print and eBook Bulk Sales web page at http://www.apress.com/bulk-sales.

Any source code or other supplementary material referenced by the author in this book is available to readers on the Github repository: https://github.com/Apress/Artificial-Intelligence-and-Human-Evolution.

Paper in this product is recyclable

Table of Contents

About the Author

Dr. Ameet Joshi received his PhD from Michigan State University in 2006.
He has over 15 years of experience in developing machine learning
algorithms in various different industrial settings including pipeline
inspection, home energy disaggregation, Microsoft Cortana Intelligence,
and business intelligence in CRM. He is currently a Data Science Manager
at Microsoft. Previously, he has worked as Machine Learning Specialist
at Belkin International and Director of Research at Microline Technology
Corp. He is a member of several technical committees, has published
in numerous conference and journal publications, and contributed to
edited books. He also has two patents and has received several industry
awards, including Senior Membership of IEEE (which only 8% of members
achieve).

Acknowledgments

The idea of writing a book in AI had been brewing in my head for a few years. I knew how I would like to start the book and what would be the central message I wanted to convey. However, I was struggling on how the book would conclude and what should be the ultimate takeaway for the readers. The availability of ChatGPT in 2022 really helped me zero in on that, and I was ready to take up the challenge.

As I had worked with Mary James, senior editor at Springer Nature, on my earlier technical books on ML and AI, I started a conversation with her. I would really like to thank her for encouraging me on this idea and connecting me with the editors of Apress. I would like to thank Shivangi Ramchandran and the whole team at Apress for helping me throughout the process of writing the book.

When I announced to my dear wife, Meghana, and two sons, Dhroov and Sushaan, that I am going to write a book on AI for general audience, they were extremely supportive. Meghana has always been a silent force behind me in my endeavors. Her unwavering support was crucial toward the completion of the book. I still remember the next day after I told this to my younger son, Sushaan. He went to his library in elementary school in search of a book on AI and did not stop till he found one. He brought it home and gave it to me with enormous pride. He told me, "Take this dad, you can copy from this and get started!" I was speechless!! My heart was filled with love and laughter at the same time. I had to explain to him that I cannot copy from an existing book and that my book has to be original, containing my own thoughts. But it was one of the most pleasurable conversations I had with Sushaan, and I cannot thank him enough for the continuous inspiration to write as well as monitoring my progress over time.

ACKNOWLEDGMENTS

I would also like to thank my father Vijay, mother Madhuri, and brother Mandar for their continued encouragement toward the completion of the book.

I would also like to thank Akshat Jaykar, for all the illustrations in this book are hand-drawn by him at the mere age of 15. As you will see throughout the book, he has done an incredible job and given a unique touch to these concepts.

Last but not least, I would like to thank my numerous friends and other family members who have given feedback on my thoughts through many discussions and helped me understand the landscape of questions our generation is seeking from this imminent new technology of artificial intelligence.

Prologue

Homo sapiens or humans are likely the first species on planet Earth who attempted to understand the glorious creation that is the universe we are part of. The incredible power, scope, complexity, and sheer beauty that the universe exudes in all the dimensions all the time just goes unnoticed. Of the 13.8 billion years in the life of the universe, the humans living on an atomic-sized planet Earth that is part of a microscopic solar system belonging to a tiny galaxy called the Milky Way are observing and appreciating and learning about it for a total of about 300,000 years! Less than a blink of an eye!! But, nevertheless, this awesome creation has received a much-awaited appreciation, and it should not go unnoticed under any circumstances!

It took billions of years for the biological life on Earth based on the carbon atoms to reach the intellectual superiority of *Homo sapiens* that enabled them to make an attempt to understand how the universe works and pave a path for dominating planet Earth while battling natural disasters and calamities. On the contrary, silicon-based life as created by humans, in the form of machines powered by computers, is already processing data about the universe to unravel its secrets the way humans could never do and is in existence for less than 100 years!!

Numbers don't lie and as the comparison speaks for itself, days of humans are numbered!! Is it true? Would silicon-based machines truly be able to lead humans to extinction? A grim picture of this sort is usually portrayed in sci-fi literature and movies and superficially it can appear plausible.

Carbon-based life was created as a result of an excruciatingly long series of naturally occurring accidental strokes in a randomly generated chemical labyrinth over billions of years through the process of natural selection. While all the silicon-based life is proactively created by carbon-based life or humans, humans have passed the test of time by surviving the calamities thrown at them for over 300,000 years and are resilient enough to survive and thrive. As such, we don't have to worry about computers taking over humans even if they are equipped with superhuman intelligence.

The first goal of this book is to learn about how we arrived at the current state, in the year 2023 AD. The sequence of events that led to the creation of life on planet Earth and its evolution from single-celled organisms all the way to the arrival of *Homo sapiens* or humans containing billions of cells. The story continues with the creation of machines by humans and imparting to them an ability to learn. We then take a moment to understand how this incredible invention is going to change human life in the near future. Then equipped with the abilities of the intelligent machines, we march on to eye for a longer-term future, where humans really take charge of their own destiny and become the species that rules the entire universe.

CHAPTER 1

Introduction

The rapid growth in the field of machine intelligence or artificial intelligence (AI) in recent years is truly mind boggling and has created an unprecedented level of excitement about the field in the minds of people who are in the know, and the people in the know are increasing exponentially. The notion of AI is not quite novel to most people in the twenty-first century, as many Hollywood blockbusters such as *2001: A Space Odyssey*, *Terminator* series, *Matrix* series, to name a few, have already introduced this concept in quite a dramatic manner starting in the mid-twentieth century. Some movies, or at least parts of them, have painted a bright future where AI is helping humans to achieve more by taking over the repetitive, boring, hazardous tasks and giving them time to focus on the activities that really matter and improve their lives, while others have painted a grim picture where AI has progressed to become a superior and more advanced species in itself and is treating humans with hostility and is taking control of the world and treating humans as slaves. Both the scenarios are interesting or exciting from the perspective of entertainment and have made the respective movies a huge box office success. However, when AI can potentially become a reality, the two implications can lead us toward very different paths and one of them is definitely not desired.

I have been meaning to write a book on the role of artificial intelligence or AI in our lives for a few years now, but narrowing down on the scope that would be apt for a single book kept me thinking and ultimately

© Ameet Joshi 2024
A. Joshi, *Artificial Intelligence and Human Evolution*,
https://doi.org/10.1007/978-1-4842-9807-7_1

procrastinating. The emergence of ChatGPT and subsequent rise of interest in the topic coupled with increasing confusion, even scare, about what AI is and what AI can do and how it is going to affect our day-to-day lives helped me zero in on what precisely I wanted to focus on for this book.

One of the crucial differences between human intelligence and AI, at least for now, is that AI is totally and completely under the control of humans. AI cannot function without humans turning the power switch ON and AI cannot reproduce itself. So even if potentially capable of doing more than humans, AI does not have free will as of today. However, the development of AI is not entirely organized or planned. Each new step or breakthrough in technology has come as a result of competition between enterprises that employ researchers and scientists and are fighting for bigger market share and higher stock prices. The next step of unlocking AI would be to let it have the two key capabilities: (1) turning itself ON and OFF and (2) being able to reproduce. If AI can acquire these capabilities in some way, it can potentially realize the ultimate fear that is depicted in the movies. Can we somehow control it so that it never happens? It can be a matter of speculation, but most likely these two options are not on the horizon for the enterprises that control the technology for now and as such are quite unlikely.

In this book we touch upon a lot of technical aspects, but this is not a technical book, it is not meant to be restricted for an audience seeking technical knowledge. It is specifically written for a general audience without any pre-requisite background in science or mathematics or computers. A reader who is interested in science or who is fascinated with the beauty of logic or is open to learning something new or is curious about the future or always likes to ask questions about anything and everything would most certainly appreciate this book. As a matter of fact, the topics in this book should be relevant to all the people on planet Earth in one way or another.

Although AI is at the heart of this book, I wanted to state at the very beginning where it all started to provide a context into where we are, how we got here, and where we are headed. I always like to compare this to the operation of bow and arrow. If you want your arrow to reach farther, you need to pull the bowstring further back. In order for us to eavesdrop into the future, we need to start investigating our past first. We start the journey with the origin of life on Earth and then follow the evolution of life forms all the way reaching up to humans. Then, we look at what makes humans the most dominating species on the planet even more so than what dinosaurs were hundreds of millions of years ago. We then look at what made humans more intelligent and superior to all the other species on Earth. Human intelligence is always the yardstick we use to measure artificial intelligence, and thus after learning about human intelligence, we enter into the era of machines and machine intelligence. We look at the present day with AI knocking on the doors with the potential to change our lives as we know it. We move past the present day into the near future with a case-by-case analysis of the impact that AI can generate on our lives without leaving the realms of reality. The last chapters of the book really dive deeper into the role and impact of AI in the near term on our lives and then follow the thread to look at a much longer-term future and how we should look at AI to help us get there.

Many predictions and theories tend to overestimate and exaggerate capabilities of AI and it makes them turn into works of fiction very quickly. AI has evolved rather too fast compared with all the other technological inventions in the past, starting with the industrial revolution in the eighteenth century. The scope of changes that took tens or even hundreds of years are likely possible in matter of months and single digit years with AI, and it definitely makes it much harder to get a realistic perspective on it. However, that is precisely the objective of this book, and I am going to make an honest and down-to-earth attempt to paint the picture of near-term future that is imminent. Being in this field almost my entire life, implementing solutions powered by AI in a multitude of fields ranging from relatively ancient oil and gas industry to present-day consumer needs on a daily basis, to state-of-the-art business intelligence and big data and search engine tech at the bleeding edge of its innovations, I have a deep understanding of what AI can and cannot do. I have seen its evolution from a disparate array of mathematical and computer science-based solutions taking bits and pieces from machine learning to its present-day human-like capabilities in the form of ChatGPT and Dall-E that can converse like a human or draw realistic images from textual descriptions, respectively.

Another relatively novel aspect a keen reader would notice throughout the book is that there are very few references despite being on the bleeding-edge topic of human evolution and AI. The only references given are about the current and historical facts; no references are needed for the concepts presented. All the concepts mentioned in the book are evolved from scratch from the first principles or common sense and there should be no need for any reader to drop this book and look up something online, unless of course they want to dive deeper into a specific topic. I like to take pride in presenting these far-reaching concepts all the way from Darwin's theory of evolution to the rise of machines and human intelligence and its comparison with machine intelligence or AI in a way that makes them accessible to general audience without a single mathematical equation or formula. Most of these concepts, especially the ones around machine learning and AI, are typically shrouded in a complex web of mathematics and computer science lingo, making them super hard for the general audience to understand and appreciate. The effects of these technologies are commonly shown through the magical things they can do in most media such as books, movies, documentaries blurring the boundaries between reality and fiction. Coupling that with the vagueness around these technologies makes it really hard for a non-technical person to truly understand what this technology is, what it can and cannot do.

This book should unlock these hidden treasures to everyone and help see, appreciate, and even predict the impact of AI just like any data scientist or researcher working at the high-tech companies would.

A big change, probably the biggest change in the history of mankind, based on the pace of it, is poised to come in the near future, and we all should be equally equipped to tackle it together to march toward a better tomorrow.

CHAPTER 2

Origin of Life

The notion of artificial intelligence, AI, or just plain intelligence at its deepest level pertains to life. There are plenty of objects in the universe as small as atoms and molecules all the way up to giant planets, stars, and galaxies, which are active in some form all the time and interact with each other, change forms, merge with each other, and even are born, but they are not necessarily alive or intelligent. As per the current theory of origin of universe, all the objects as well as source of energy came into existence starting with the Big Bang around 13.8 billion years ago, and since then, all these entities have been at work at every instant of time. Why and how exactly the Big Bang was initiated or what was the universe made up of before the Big Bang is something we don't know and may never know with full confidence. However, if we assume that something made it happen, we can understand, or at least estimate, all the subsequent activities. What is more important is that we can understand the current state of the universe with a lot more detail and confidence. We understand that our Earth is a planet, and it revolves around the Sun along with seven other planets. We understand how the orbits of each of these planets and their speeds of rotations around themselves and around the Sun are determined by their individual masses and the mutual distances between them and the Sun as well as between each other. We can calculate these metrics based on Newton's law of gravitational force (or even more accurately using Einstein's theory of general relativity and curvature of spacetime). We even understand to some extent how the Sun itself is bound with the center of

© Ameet Joshi 2024
A. Joshi, *Artificial Intelligence and Human Evolution*,
https://doi.org/10.1007/978-1-4842-9807-7_2

our galaxy, the Milky Way, and rotates around it and how the Milky Way galaxy itself interacts with our neighboring galaxy Andromeda, and so on. We can also observe and predict the motions of various asteroids and comets based on their sizes and distances. However, from the perspective of being intelligent, none of these objects really exhibit intelligence. These objects are just following the fundamental laws of physics, and there is nothing new coming out of them. Even though at the very microscopic level, quantum mechanical laws may indicate some random and unpredictable events; at the macro level, none of these objects exhibit any unpredictable behavior or a behavior that would proactively change their actions for achieving some ulterior goals. For example, we don't see planet Mars proactively trying to come close to the Sun by altering its shape in order to get more heat, as it's feeling cold; or we don't see an asteroid changing its direction to avoid a crash with planet Earth, which is bound to terminate the asteroid permanently. This is where we draw a line and separate the living from the non-living, separate intelligent objects from not-intelligent objects. In summary, the word intelligence, the way we interpret it at least, is intrinsically linked with the concept of life. Hence, we will start our journey with the notion of life. In this chapter, we will explore what life is and what really separates it from everything else.

What Is Life?

"What is life?" It is one of the quintessential questions in the fields of philosophy as well as science. Even if the very definition of life can be daunting, we are mostly very good at classifying any object as alive or not. For example, we know that rocks are not alive, water is not alive, air is not alive, but dogs are alive, birds are alive, even fish are alive. When it comes to trees, it becomes a little tricky as they don't move around like most other living things, but we still somehow know that they are different from rocks and rivers and are alive as well. So, what is it that separates the life-forms from not life-forms?

The comparison between different versions of philosophical definitions of life can be a topic broad enough to deserve a book in itself. We will rather focus on scientific or logical interpretations of life in this book. One of the most widely accepted explanations of life as worded in various encyclopedias and dictionaries identify life as something that encompasses biological processes and some form of signaling and self-sustenance, capacity of reproduction, energy transformation, and growth. This definition can appear too complex and, as a result, far-fetched even by some loose standards. The idea that is trying to be captured here is a list of parameters that can distinguish the living from the non-living. It seems to be hinting at something like a Venn diagram,[1] where the explanation is trying to draw overlapping circles to isolate everything that we see around us that is living from what we think that is non-living. Let's consider each of the properties that are listed in the definition one by one and to draw these circles to create the Venn diagram.

Let's start from the last property; Growth: Growth is characterized by physical increase in volume and mass as well as change in shape. Let's try to apply this concept to various objects we see around us. We obviously know that all living organisms show growth, but what about some non-living organisms, say rocks? We can see that some rocks can grow as they roll with the wind and more soil sticks to them or as they move with streams of water and merge with other rocks, essentially showing signs of growth. What about icecaps on top of the mountains? They grow each winter with snowfalls. Are these objects living? We don't think so! Thus, if we have growth as the single rule to separate living from non-living, it is not sufficient. So, we need more rules or more circles. Let's start adding more of them to the Venn diagram shown in Figure 2-1.

---

[1] A Venn diagram is a diagrammatic representation of sets. It uses overlapping circles or other shapes to illustrate the logical relationships between two or more sets of items. Each circle represents a set, and the overlapping areas represent common elements of the sets.

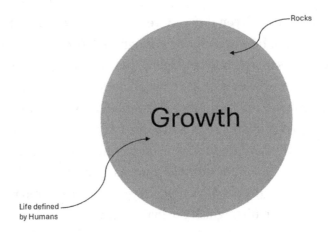

Figure 2-1. *Venn diagram with one of the parameters from definition of life*

Let's take up the second one from the definition, the "energy transformation." Once again, we know that all living organisms are capable of consuming some form of energy as required for their life supporting activities, thereby transforming one form of energy into another. Looking back at rocks and icecaps, we don't see clear indications that rocks and icecaps can proactively transform energy from one form into another by themselves. Thus, we can indeed separate them from living things. But what about some other non-living objects such as rivers? We can see that rivers flow from one point to another. With them, they carry pebbles, soil, and rocks from one point to another. As these objects flow with water, they generate heat by friction. Thus, they exhibit a definitive capacity to convert kinetic energy into heat energy. Along with energy transformation, rivers can grow in size as well each year during rains. Now, are rivers alive then? We still think not!! So these two circles, or rules as shown in Figure 2-2, are not quite sufficient to separate living and non-living and we need to proceed with one more. We can continue with our Venn diagram by adding one more rule from our earlier definition.

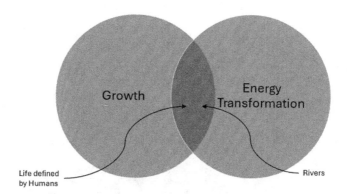

Figure 2-2. *Venn diagram with two parameters from the definition of life*

Let's add capacity to reproduce, as shown in Figure 2-3. We are quite certain that all the living organisms, from single celled ones to complex mammals, can reproduce. But what about rivers? Typically, the concept of reproduction is defined in the context of biology, and therefore, cannot be applied to non-living things directly. However, we can take some literary and philosophical liberty and use a more general and broader definition of reproduction as the ability of an entity to create a new entity that resembles the original entity. With this new interpretation or reproduction, let's revisit rivers which seem to satisfy the earlier two conditions. When a river comes across a hill or some obstacle, it can separate into two or more streams. Each of these smaller streams possesses all the similar properties of the original river, and can be considered as offsprings of the river, essentially proving that rivers can indeed reproduce. Thus, with our broader definition of reproduction, rivers can be considered as living. But are they really? We still think no! We need to proceed with our definition and add more rules or circles to our Venn diagram.

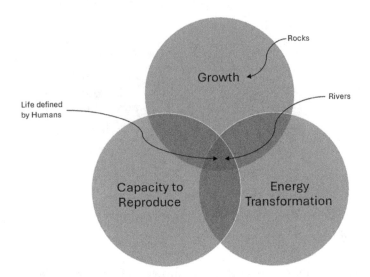

Figure 2-3. *Venn diagram with three parameters from the definition of life*

Now, things start to get more interesting as we start adding more specialized biological concepts to the mix. Self-sustenance by definition states that the entity can sustain itself without needing external help. It can be interpreted in multiple ways and with most general definition, any living or non-living object can be considered as self-sustaining. For example, rivers can continue to be rivers as long as there is a continuous supply of water at its origin. If the source of water is somehow depleted or evaporated, the river would cease to exist. But similar can be stated in the case of living things also. If a tiger does not get anything to eat for a week, it will perish and will cease to exist. In contrast, a river may exist much longer. Thus, self-sustenance, along with all the previous rules, is still not sufficient to separate living from non-living, as shown in Figure 2-4. So, we continue with our definition to add more rules or circles to see if we can classify river as non-living in the Venn diagram.

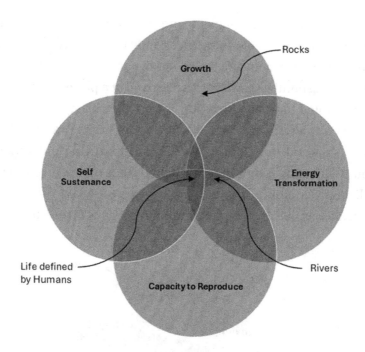

Figure 2-4. *Venn diagram with four parameters from the definition of life*

The next one is the ability to perform some form of signaling, as shown in Figure 2-5. This is an interesting one. In living organisms, the cells can signal to other cells to transmit/receive messages or information. Do we see such signaling in the case of rivers? Not in most situations. On some rare occasions, we may see lightning strike a river from clouds in the sky, potentially transferring some form of electric signal from the clouds to the river. Can that be considered as signaling? The answer to this question is negative. There is an intrinsic assumption of proactively sending a piece of information from the sending party and understanding and reciprocating the signals as they are received by the receiving party as per the definition of signaling. Are clouds really sending any information through lightning here? And even if they are, can rivers understand those signals and reciprocate by sending back a signal to the clouds? Definitely not. Thus, this random occurrence of lightning is certainly not a form of signaling. We can then move rivers outside the center of our Venn diagram that is reserved for living entities. Have we then arrived at a place where we have completely isolated the living from non-living? We will need to consider a few more examples. How about cellphone towers? They are man-made and not natural, but nonetheless they are non-living and they are continuously signaling with hundreds or thousands of devices all the time. Are they living? Certainly not, and it is rather easy to separate them using the earlier rules, as these towers do not show any signs of growth or cannot reproduce.

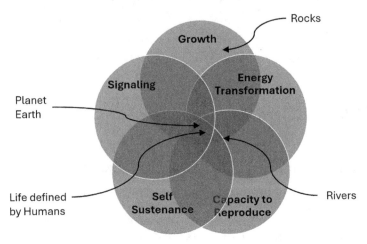

Figure 2-5. *Venn diagram with five parameters from the definition of life*

At this point, we seem to eliminate most entities that we believe to be non-living. Is our Venn diagram final then? Not quite! Let's move outside of Earth and look at some of the bigger celestial objects. How about planets, heck, how about our own planet Earth, for example? Let's see how it fits in the Venn diagram. We can consider that Earth to be continuously signaling with the Moon to keep it in the orbit through the power of gravity and Moon reciprocates by staying in the orbit. Earth can certainly grow in size, the way it has since its formation with our Solar System, when it got hit with millions of asteroids and comets, absorbing their mass. Earth is definitely capable of self-sustaining. Earth can also potentially reproduce to create a smaller planet-like object if it gets hit with a really big asteroid that breaks out a small mass from Earth. Some theories already suggest that our current Moon was created in the same way from the Earth already! Also, millions of energy transformations are happening continuously on Earth. Seems like Earth can lie at the center of all the circles along with all the living organisms. So, is Earth a living object? Well, No!! Definitely not!! This leads us to the final circle in the Venn diagram, as shown in Figure 2-6!

Biological processes: A biological process is defined as a series of events or actions that occur in living organisms to achieve a particular result. These processes are essential for the organism to grow, reproduce, maintain homeostasis (a stable internal environment), and respond to their environment. Essentially, biological processes are the ways in which an organism's cells work together to support life. This is the final circle where we come up with something that would eliminate everything in the universe that we think is not living! One might say that this rule involves a circular argument,[2] where we are essentially stating that non-living objects are things that do not possess properties of living objects due to the presence of the word "biological." However, that is not necessarily true. All the biological processes can be perfectly and completely defined using purely chemical and physical rules and laws with logical statements that apply equally well to living vs. non-living objects without explicitly using the notion of life or biology for that matter.

In spite of being host to all the living creatures, the Earth itself is primarily made up of rocks and minerals, and water and does not

---

[2]A circular argument, also known as circular reasoning or begging the question, is a logical fallacy where the conclusion of an argument is used as a premise of that same argument. Essentially, it's when someone tries to prove a point by using what they are trying to prove as the evidence. One can prove anything with circular reasoning.

contain cell structure in itself and does not show the aforementioned biological processes. Hence, Earth as its own object does not satisfy this condition. With this final rule, we arrive at the Venn diagram where we can isolate all the living objects from the non-living ones as shown in the accompanying figure.

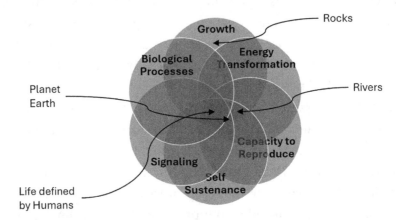

Figure 2-6. *Final Venn diagram with all six parameters. Human-defined life now emerges as the only item at the center overlapped by all the circles*

Thus, we have completely isolated the human-defined life forms from all the other entities in the universe. If we look a bit deeper, one can argue that with the last rule in the Venn diagram, do we really need all the other rules? It might appear that this single rule can potentially separate the living from non-living. Then why do we need to include all of them? The answer to this question lies in the distinction between the living and the dead. The dead organisms do possess all the biological processes and cellular structure (at least for some time till all the tissues and cells have disintegrated), but they are not "functioning." They are not signaling or moving or are able to reproduce and so on. Thus, we essentially need all the other parameters to distinguish between the living and the dead, but only the first parameter is sufficient to distinguish between the living and the non-living.

How Did Life Originate?

Now, equipped with the full definition of life and what truly separates the living from the non-living and dead, let's look at the question of how it all started. Did life start on Earth, or did it come to Earth from some other planet, or asteroid, or a star even? Based on the atmospheric and thermal conditions required for most species on Earth, it is rather difficult to imagine a life-form coming to Earth from outer space. But is it truly impossible? As it appears, it is not!!

The idea of an alien life is not new, and it forms an entire genre of literature in science-fiction. But with some recent developments in the field, this idea has been brought close to reality and is not just science-fiction anymore. Let's take a closer look at the option of life on Earth originating from outer space.

Extremophiles

It is quite clear that all the "large species" or the species that we can see without needing a microscope cannot survive outside of Earth for longer than a few minutes, if that. The reasons being lack of oxygen, extreme cold, excessive radiation from the Sun and/or other stars, lack of gravity, etc. All the large species would die within a short time if either of these factors were encountered. When all these factors come together, there is just no realistic chance for survival for these species. However, scientists have found a few types of single celled organisms on the Earth that can potentially withstand extreme temperature, pressure, or atmospheric conditions and they are called extremophiles. These microorganisms can survive in places like hydrothermal vents deep inside oceans, where temperatures can rise as high as 120° Celsius or 250° Fahrenheit along with very high pressure. There are also some microorganisms that can sustain extremely cold temperatures such as –25° Celsius or –13° Fahrenheit buried inside of ice. There are some microorganisms that can sustain extreme acidic or alkaline conditions or very high salinity. There also exist some microorganisms that can live entirely in the dark as well as there are microorganisms that can sustain dangerous radiation such as cosmic rays or X-rays and radiation from radioactive materials. Any of these conditions even in moderation would be sufficient to kill the entire human race, let alone other animals. However, these microorganisms can not only survive in such extreme conditions but also can grow and reproduce in such conditions.

It is quite encouraging that these microorganisms can survive in these harsh conditions on Earth, but what about outside of Earth? Can they take these capabilities and survive in outer space? Where there is no sunlight, no gravity, and continuous, unhindered cosmic radiation from all the neighboring stars. This cosmic radiation is reduced to only trace amounts on the surface of the Earth due to absorption by Earth's atmosphere and deflection by the Earth's magnetic field.

To answer this question, a research team from Japan's Kibo lab and Japan's Space Exploration Agency (JAXA)[3] conducted an experiment on the International Space Station. Their experiment included multiple different species of bacteria, but specifically, the bacterium of type *Deinococcus radiodurans* stood out as the most resilient and was able to survive for multiple years in the space outside the International Space Station. When a group of surviving bacteria were examined, they observed that the cells in the outer exposed region had died, and the dead cells created a sort of a protective layer, under which the inner cells were able to survive. This research conclusively proves that there is a definitive possibility that microorganisms can sustain the harshness of outer space for a long time and can still continue to live in a dormant condition.

---

[3] Smithsonian Magazine: [www.smithsonianmag.com/science-nature/scientists-discover-exposed-bacteria-can-survive-space-years-180975660/]

However, from the perspective of finding a possibility that life on Earth could have come from outer space, sustaining this harshness of space is only one part of the equation. The other parts include the ability to sustain the impact when the meteorite or comet carrying these bacteria hits the surface of Earth, and the ability to use the energy from the Sun directly or indirectly to grow and reproduce. Specific experiments to see if *Deinococcus radiodurans* can survive such an impact have not been carried out and currently, the answer is more of a speculatory nature. *Deinococcus radiodurans* bacteria need oxygen and the presence of organic compounds to generate energy for growth and reproduction. They do not possess the capability to directly consume Solar energy and hence cannot be classified as a completely independent species. However, the emergence of such species provides a path where it is possible for yet another type of species to have evolved on another planet that can take a step further and be able to survive the impact as well as be self-sufficient in consuming energy from Sun directly along with having the ability to stay alive in the space for many years.

Even if we agree with this theory of life's origin outside of Earth, the fact still remains that the species that came to Earth from outer space must have originated somewhere else. May be a planet in some other star system in our galaxy or even from another galaxy. But then another question inexorably pops up: how did life originate on that planet? So, until we answer the ultimate question of origin of life from non-living substances, the chain of questions does not really end.

Let us now look at the other option where life originated entirely from scratch on Earth! This option is also typically a more favored option.

Let's try to explore how it could have happened. Our solar system came into existence around 4.6 billion years ago, when the Sun was first formed. It took another 100 million years for the other planets to form, completing the solar system. There are many theories that have been proposed to explain the formation of our solar system. They differ in many aspects, but most of them agree that it all started from a giant cloud of gas called the solar nebula. It was quite cold and contained mostly hydrogen along with some heavier elements in minuscule amount. Either on its own or with the help of a perturbation likely from a distant supernova explosion, a portion of gas clumped together to form a seed for the formation of star. Once the mass started to concentrate in the clumped region, the increased gravitational force of that region accelerated its growth, and the region quickly grew to become the most dominant entity in the nebula. With the increase in size and corresponding gravity, the center of the clump started to get warmer, ultimately reaching temperatures high enough (about 100 million degrees Celsius) to start the process of nuclear fusion. Thus, our Sun was first formed. After the Sun's formation, it quickly absorbed all the light gases around it leaving only heavy elements in the close neighborhood. These heavy elements clumped together over the next 100 million years to form the first four planets in the forms of Mercury, Venus, Earth, and Mars. At a distance far enough from the Sun around the current location of Jupiter, the Sun's gravitational pull was not as strong and light gases could still exist. They started clumping together to form the gas giant planets like Jupiter and Saturn.

Origin of Solar System

Focusing our attention on Earth's formation, the first iteration of Earth primarily consisted of the heavier elements that were left out of the Sun's gravitational force such as Carbon, Nitrogen, Oxygen, and heavier metals, etc. With the increasing size of the Earth, it started to get its own gravitational force and with that, its own unique identity in the Solar system. This period extended for about 400 million years from 4.5 billion years ago to 4.1 billion years ago. This period is called the Hadean period and the Earth started to cool down after its formation during this period. This period was followed by what is typically called as a period of Late Heavy Bombardment (LHB) period. The LHB period started around 4.1

billion years ago and lasted for another 300 million years till about 3.8 billion years ago. During this period, the formation of bigger planets like Jupiter, Saturn, Uranus, and Neptune and their increased gravitational pull created a disturbance in the asteroid belt that was present on the outskirts of the Solar system. This disturbance initiated the heavy bombardment from meteorites, comets, and other space debris. Although quite violent, this was an extremely crucial period from the perspective of the origin of life on Earth. These meteorites and comets brought key ingredients to Earth, including water. Due to this continuous bombardment, the Earth's temperature rose again to higher than the boiling point of water and quite a significant amount of water must have been lost as well. However, whatever remained was sufficient for the formation of life. It is important to note that all the neighboring planets, including Mercury, Venus, and Mars, also encountered a similar bombardment and likely received a similar dose of meteorites and comets during this period. However, due to various factors, such as proximity to the Sun, lower gravity, etc., they lost most of the water and other elements in due course. However, Mars has a very close geological structure to Earth and similar size, and it could have retained water and other elements for a longer time than Mercury and Venus and there is a reasonable possibility that life could have independently originated there as well. However, we have not found conclusive evidence of it yet.

Around the end of the LHB period, or around 3.8 billion years ago, the Earth started to cool down again. As per National Science Foundation, some of the oldest rocks discovered on the Earth near Hudson Bay in Canada date back to this time.[4] However, we don't have any real evidence of what existed on Earth before this time.

There are mixed opinions on exactly when life originated on Earth. Some theories suggest it could have started as early as 4.3 billion years ago, even before the LHB period itself, while other theories suggest it

---

[4]www.nsf.gov/news/news_images.jsp?cntn_id=112299&org=NSF

started during the LHB period or around 4.1 billion years ago. However, the oldest fossils found so far, that can prove the existence of life, only date back to the LHB period and, as a result, there is no realisleic way to prove or disprove the other theories.

Primordial Soup

As per the early discussion on definition of life, in order to prove that life originated on Earth, we need to prove the creation of biological structures capable of displaying functions such as growth, signaling, reproduction, etc. from entirely non-biological substances. There are many theories that attempt to provide mechanisms of how this could have happened. All these theories start with a mixture of primitive organic compounds containing Carbon, Nitrogen, Hydrogen, and Oxygen such as Methane and Ammonia with water and dissolved Hydrogen and Oxygen, typically called a *Primordial Soup*. The word primordial means something that existed since the beginning of everything. In this case, primordial soup means a mixture of all the organic compounds that were available on Earth naturally that led to the creation of life. All the different theories that base the origins of life on Earth on the chemical reactions that naturally happened are referred to as primordial soup theories.

One of the most compelling primordial soup theories is called as *RNA World Hypothesis*. As per the RNA World Hypothesis theory, the first living microorganisms were entirely based of RNA structures. RNAs are the smallest entities that can perform most biological functions such as signaling, reproduction, growth, etc. A single RNA molecule is the most basic part of living cells and contains a single strand of long chain of atoms arranged as a helical structure. Carbon atoms form the central part of the chain in the RNA molecule along with atoms of Hydrogen, Oxygen, Nitrogen, and Phosphorus. The DNA or Deoxyribonucleic Acid found in most living organisms is a dual stranded molecule where its single strand is similar to an RNA. DNAs typically have a double helical structure.

Most of the living cells that we observe today are based on DNA molecules, while only some viruses contain RNAs. This observation would make it mandatory that DNA molecules should have formed naturally from some chemical reactions. DNA molecules are significantly more complex than RNA molecules and it is far less likely that DNA molecules can form with random chemical reactions. However, a recent study published in Nature[5] magazine supports the theory that early life that originated on Earth can be entirely based on RNAs.

The long chains of atoms in the RNA molecules can be broken down into four distinct groups of atoms called adenine, guanine, cytosine, and uracil. (DNA molecules share three of these groups, with the difference of uracil being replaced with thymine.) These compounds are called nitrogen bases or nucleotide bases. Unique properties of RNA molecules include the ability to encode information in the form of sequence of the atoms in the chain and ability to act as a catalyst to initiate biological processes. It has also been shown that RNAs are capable of reproducing and transmitting the information contained in them to the offsprings.

About 4.3 billion years ago, it is believed that environmental conditions on Earth reached a critical stage where all the necessary requirements for inducing the chemical reactions for generation of RNA type molecules were available. These conditions include (1) Formation of Earth's crust. The reverberations from all the early bombardments of meteorites had settled and the surface of Earth began to solidify, forming what we call the Earth's crust. (2) Moderate temperatures on the surface of Earth. The average temperature on Earth at this time dropped below 100° Celsius, making liquid water possible. It could have been spanning from subzero temperatures around poles, but overall, it was conducive for the chemical reactions necessary for origin of life. (3) Formation of liquid water. The moderate temperatures paved the way for the existence of liquid water around this time. The water accumulated during the LHB period was

---

[5] www.nature.com/articles/d41586-022-01303-z

sufficient to create oceans covering over 70% of Earth's surface. (4) Presence of complex organic compounds. Earth contained significant amounts of Carbon, Hydrogen, Nitrogen, and Oxygen, along with trace amounts of Sulphur and Phosphorus. (5) And last, but not least, the dense atmosphere on the Earth was able to cause heavy lightning. Such lightning strikes were able to trigger chemical reactions with the complex organic compounds producing RNA type molecules.

The Miller-Urey experiment provided crucial evidence for this process. In 1953, two professors at the University of Chicago, Stanley Miller and Harold Urey, conducted this experiment. The experimental setup was quite simple. They brought two complex organic compounds, Methane and Ammonia along with water and hydrogen gas and sealed them together in a flask. It has been shown that such compounds can be easily produced when the underlying elements are exposed to UV radiation. Then they connected a pair of electrodes to the flask and applied electric sparks repeatedly. Over just a few hours of operation, it was observed that multiple amino acid molecules were formed in the flask. These amino acid molecules are the building blocks for the nucleotide bases that ultimately form the RNA and DNA molecules.

To further advance this concept, Thomas Carell from Ludwig Maximillian University[6] based in Germany has recently demonstrated a sequence of chemical processes that conceptually can make not just the amino acids but all the four nucleotide bases that are the building blocks of an RNA (adenine, uracil, cytosine, and guanine) from basic ingredients such as water and nitrogen under atmospheric conditions readily available on Earth around 4.3 billion years ago. With these raw ingredients for the construction of RNA available, over the next several thousand or million years, they were grouped together to form the first RNA molecule.

There are some other theories that propose life originated deep underwater. Specifically, near the underground vents at the bottom

---

[6] www.nature.com/articles/d41586-019-02622-4

of the oceans there lies an interesting mix of ingredients such as high temperature to initiate chemical reactions, presence of necessary elements through the fumes coming out of the vents and liquid water. Even if sunlight is not present at such depths under the ocean, the heat coming from the vents is sufficient to provide the necessary energy for life to survive and grow. The idea is that RNA type molecules were first produced around these vents and over thousands to millions of years, they kept combining to form microorganisms with cellular structures. Some of the earliest organisms known to man do come from under water, and this fact favors this theory of origin of life. However, there is not enough experimental backing for this theory similar to the one provided by the Miller-Urey experiment or Carell's experiment for the theory of origin of life on the land.

There are some other theories that claim life could have originated under deep sheets of ice, or inside clay or soil, etc. But most of these theories lack sufficient experimental backing and are not favored by many scientists.

Ultimately, last, but not least, the discussion on origin of life is never complete unless we consider the involvement of some superior power such as God. When one claims that the life on Earth or in the Universe was created by God, there are not many arguments that can be made against it if we go in the direction of faith. However, we can certainly ask some probing questions if we take this statement in the spirit of logic. If God or the superior entity did create the life, why did it create from the small microorganisms? Why did it not create humans or *Homo sapiens* directly? Why did it have to go through the long process of evolution? If God created life on Earth, what else did God create? Why was life created only on planet Earth and not on any other planets in the Solar system? If God created life on Earth, God must have created the Solar System also, and extending the argument, the entire universe as well. If God created the entire universe, why did he/she wait for billions of years to create the Earth and life after creation of the universe? There is simply no limit to what questions we can ask, and there is no good answer to those other than God wanted to do it that way. So, if God wanted to do all these things in a certain way, in a certain sequence, in a way that follows well-defined laws of Physics and Chemistry, then we can still try to study our understanding of these laws, improve the equations based on the evidence we see and be able to predict how we came here and where we will go next. Thus, either way, if God created us or not, we still need to do our due diligence and learn about the universe, learn about our solar system, learn about the Earth, and find out how it all started!

Conclusion

Life, as we all know it, indeed differs from other concepts in the universe. However, its objective definition may not be obvious as we saw in this chapter. We take life for granted as we see so many different life forms around us, including us, but at the start, it was far from obvious. It was rather one of the unlikeliest things to happen statistically speaking. However, it was not impossible and happen it did. Is the primordial soup the true answer? May be, may be not! But definitely some chemical reactions happened using atoms and molecules that were readily available on Earth that led to the formation of the first biological structure. In the next chapter, we will look at how life progressed through time.

CHAPTER 3

Evolution of Species

Introduction

Our perception of intelligence is tightly integrated with life and living organisms. In the previous chapter, we looked at various processes that led to the origin of life on Earth in the form of single-celled organisms. Routine cellular operations do not show much in the form of intelligence. It is only when we start looking at the actions of complex multicellular organisms such as animals and birds and even aquatic species that intelligent behavior starts to surface. It is not just a coincidence, and intelligence is indeed deeply rooted in the biological structure of the species that exhibit it. Hence, in order to fully appreciate the concept of intelligence, we need to learn about these organisms and how they came into existence. In this chapter, we will dive deeper into the topic of life and look at how the life that originated on the Earth culminated into the creation of complex species such as humans containing trillions of cells.

Evolution of life

The first organisms that were created on Earth were primitive single-celled organisms based on a single RNA or DNA and were capable of carrying out all the fundamental life processes. However, the species that we see

© Ameet Joshi 2024
A. Joshi, *Artificial Intelligence and Human Evolution*,
https://doi.org/10.1007/978-1-4842-9807-7_3

around us today, including us, are far more complex and contain not just hundreds or thousands, but billions and trillions or even quadrillion cells, where the same fundamental life processes exhibited by the single-celled organisms are now handled by dedicated organs that contain super-specialized cells. Transformation of the primitive single-celled organisms into these highly advanced species did not happen in a short time, or by accident. It took billions of years to reach this level of complexity. There is sufficient archaeological proof that the process that was responsible for it is *Evolution.*

The theory of evolution by natural selection is credited to Charles Darwin. Darwin was one of the greatest scientific minds that is known to us. Darwin proposed his theory of biological evolution in the year 1859 AD in his coveted book *On the Origin of Species.* Theory of evolution is one of the most elegant and ground-breaking theories ever proposed in the history of science, let alone biology. Although it is based on assumptions that are simple enough for an elementary grade student to understand and appreciate; a repetitive chained application of those principles is capable of explaining births of the complex web of all the species that have ever lived on the face of Earth, starting from single-celled organisms. Being fundamentally simple is a quintessential litmus test for many great theories. When a theory is simple, it can either be easily refuted or is nearly impossible to refute; there is little left for interpretation; there is no obscurity in it. There is no vagueness caused by a long and complex chain of deductive arguments. When a theory gets complex, there are typically numerous cases to consider and although they can be hard to prove, they are also hard to disprove. In the worst case, a complex theory can hide behind its vagueness, till enough research proves all the cases, or in the best case, the theory becomes limited in accessibility only to people that are capable of understanding the full complexity of it. A classic example of such complex theory is *string theory* or its generalization in the form of *super-string theory* or *M theory.*[1] Also, when a theory gets more and more complicated, the scope in which it is applicable starts to reduce. In other words, theories that are simple in nature are rather broadly applicable, while complex theories try to explain some very specific scenarios with significantly narrower scope. Having said that, it takes a genius to propose or develop a theory as simple as evolution and be valid and capable of changing our understanding of life on Earth.

Sometimes, a theory in concept can be *relatively* (pun intended) simple, but its formal definition can be mind-bogglingly complex. An example being Einstein's general theory of relativity that tries to generalize the Newtonian physics, especially the concept of gravity. It defines gravity not as a force like Newton did, but as a curvature in the space–time fabric. Thinking of curvature in space–time is not too hard to understand, especially when it is dialed down to 2 dimensions. A simple two-dimensional graphic of a checkerboard type surface getting distorted when a large object appears on it as one can see in Wikipedia and many other sources is a great example. However, to understand the full mathematical formulation of it, using tensor calculus, earning a PhD becomes only a starting point. It is also important to note that the real differentiations in visible implications of general theory of relativity compared to Newtonian Physics only come into the picture when we go beyond Earth, and even our solar system and look at stars and black holes that are light years away; or we need to look at some minuscule quirks in the orbits of planets that only a handful of telescopes on Earth can measure. Newton's theory of gravity and motion, which can now be looked upon as a simplified version or a special case of the general theory of relativity, can answer most questions in real life. As hard as it may be to believe, all the sophisticated architectures and engineering marvels on Earth over the last three to

---

[1] Although the string theory and its variants try to explain and unite all the physical processes into a comprehensive framework, it still remains incomplete and experimentally unproven.

four centuries are built using Newton's laws and not general theory of relativity. Crystal Palace, built in 1851 in London, was one of the first largest constructions at the time built making exhaustive use of Newton's laws. It was constructed using glass and iron structure to make it strong but light and it was an absolutely incredible engineering feat at the time. The success of Newton's laws just kept climbing up and up with breathtaking constructions of the Eiffel tower in Paris, the Brooklyn Bridge in New York, USA, all the way to the Burj Khalifa in the United Arab Emirates, which is the tallest building on Earth by a significant margin (about 200 meters or 600 feet taller than the runner up), to name a few. Even the design of the gigantic architecture of Burj Khalifa did not need help from general theory of relativity, Newton's laws were accurate enough.

This is not to discredit the greatness of general theory of relativity in any way, which truly marks the epitome of modern-day physics, mathematics, and astronomy in the twentieth century; but it does illustrate a key aspect of scientific theories that complexity of solution tends to follow the complexity of the problem.

The theory of evolution is possibly even simpler to understand than Newton's theory of motion and gravity and needs no complex augmentation whatsoever to make it more general (as yet!). This is a work of pure genius!! In order to move forward with the specific details of human evolution, let's quickly go over to the fundamental principles of Darwin's theory of evolution. Darwin's theory of evolution is based on two steps that take place in a repeated fashion generation after generation in the lifecycle of the species:

1. Random genetic variation in offsprings by the process of mutation during reproduction

2. Natural selection dictated by survival of the fittest

Let's look at the first aspect of the theory. Every time a species reproduces, an offspring is generated by mixing the genes of the parent(s) with some degree of random variation. If such slight random variation did not exist, then the offsprings would exhibit exactly the same traits as their parents, and no new features would ever be generated. At the same time, if the offsprings are generated using excessive random variations, the continuity of the species would be lost and whole new species would be created in every generation. Thus, there needs to be a delicate balance between randomness and predictable identical replication. These random variations can also be looked upon as errors or accidents in the creation of offsprings when ideally there should not be any variation. With such carefully balanced randomness or accidents, each reproduction leads to generation of slightly different offsprings. The variations in the genetic structure can happen due to mutations, genetic recombination, or even environmental factors. Mutations are essentially the innate errors or accidents in DNA replication as mentioned earlier and can be caused by environmental factors such as exposure to radiation or some harmful chemicals or due to a viral infection. Mutations are commonly observed when cell division happens through the process of meiosis. Genetic recombination is another process by which new combinations of DNA sequences can be created. During meiosis, the chromosomes or DNA sequences in a cell are shuffled, and new combinations of genes are created. This can also lead to new variations in the population. The continuous mutations across all species make the habitat a vibrant and dynamic place where mutual interactions between species are changing continuously.

This dynamic habitat sets the stage for the second step in the process of evolution: Natural Selection. The newly formed offsprings then interact with the environment with three possible outcomes: the new features produced as a result of the mutation make them better at surviving the

environment or they make them worse at surviving the environment
or they don't really make any measurable difference. The third case
essentially is a scratch and makes no real impact. In the second case,
statistically speaking, such offsprings tend to die sooner with less chance of
reproducing and passing those inferior features to their next generations.
The offsprings with features that make them better suitable at surviving
the environment tend to live longer, especially in contrast with the species
with inferior features, giving them a higher chance of reproducing and
preserving those new features through generations. It should be noted
that these changes cannot be clearly perceived on an individual basis.
They should be looked at from the context of large sample space across
the entire habitat and through the lens of statistics. Over hundreds or
thousands of generations, a definitive drift toward the features that make
the species better and better at surviving the environment becomes
apparent. It is also important to note that the environment as such should
stay fairly consistent in order to see this effect. When the environment itself
starts to change rapidly, it may not give enough chance for the species
to evolve in any specific direction to sustain the changes and may lead
to mass extinction. However, when the changes in the environment are
slow enough, the slow drift toward new and improved features leads to a
generation of a whole new species over time that are better suited to the
changing conditions. Rapidly changing environment, even when it is not
completely devastating, can still make the process of evolution ineffective
by resetting it too often. It is important to note that there is always a race
for survival among different types of species and each of them is evolving
simultaneously. As a result, the whole process is quite dynamic, but there
is an inherent and delicate balance. However, eventually the statistics
tend to favor one species over another, and that species evolves into
becoming a dominating one. A great example of such a phenomenon was
the emergence of dinosaurs. Their eating habits ranged from carnivores
to herbivores and their sizes ranged from measly two to three inches to
nearly 100 feet. However, they all had one thing in common: they all were

so much better at surviving in their respective environments that history saw an almost unhindered growth of them across the planet. However, one day, in a fell swoop, they were completely eradicated from Earth with a catastrophic calamity, and it showed that even the most evolved species can perish to natural disasters when their impact changes the environment too fast. During this catastrophic event, although dinosaurs perished, some smaller animals and plants did survive and were able to get on a brand-new evolution train once again with the newly formed environment and ultimately culminating in the emergence of humans.

Adaptation

The concept of *adaptation* describes a process of change by which a species becomes better suited to its environment. Adaptation is quite intricately linked with the process of natural selection, and hence evolution in general, but they are fundamentally different processes. Evolution takes multiple generations, ranging from hundreds to thousands, to create new species by altering the genetic structure that is better suited to the given environment as discussed earlier. The rate of incremental change through each generation depends on many factors such as rate of mutation, size of population, and the environment itself. While adaptation is a change in an organism without changing the genetic makeup to help it better survive the environment, adaptation shows its effect on single or few organisms and can happen at a much faster rate, while evolution changes the entire population.

Here are some of the examples that are generally given to illustrate the process of adaptation: evolution of polar bears with white colored fur to better hide in snow, or webbed feet of ducks to make them better swimmers, sharp teeth of carnivores, large beaks of birds, whiskers, claws, and the list goes on and on. It is important to note that not all of these adaptations have happened in a few generations, as some of them took over hundreds of thousands of years and likely over corresponding number of generations.

As a matter of fact, an implicit rule that an organism cannot adapt to its environment in a single generation is at the very heart of Darwinian theory of evolution. If that could happen, then the need to change the genetic information in the form of mutations can become optional. The organism is born with a fixed set of features and is bound to react with the environment with the help of those features and those features alone. There is nothing that can be changed. The only change that can happen is when the organism reproduces and mutation kicks in. However, the changes are only seen in the offspring. If an organism could adapt after birth, the process of natural selection needs a whole new interpretation. Typically, all the adaptations that are observed can be classified into two types:

1. Physical adaptations

2. Behavioral adaptations

Physical adaptations describe the changes in the body of the organism. The changes like development of wings, webbed feet, sharp claws capability of running faster, changes in height/weight, capability of breathing under water, etc. are all examples of physical adaptations. Obviously, it takes thousands of years to see these physical changes, and in some cases, they lead to the creation of a whole new species, thereby merging with the process of evolution.

Behavioral adaptations describe the changes in the behavior of the organisms, like hunting in a group, living in a colony, courtships for finding mates, etc. It is important to note that even if behavioral adaptations don't necessarily need changes in physical appearance, they are not seen to happen in a single generation. Even if these adaptations don't need any changes into the physical properties of the species, they are related to the structure and functioning of the brain and nervous system and the organisms are born with a pre-programmed brain to show these behaviors. These behaviors also do not change with the changes in environment automatically.

Starting with Single Cell

The principles of random variation of features in offspring during reproduction and natural selection coupled with adaptation in the Darwin's theory of evolution do provide an architecture that can explain the occurrence of modern humans starting from single-celled organisms about 4 billion years ago, but it does not provide direct explanation of everything that has happened with life in the last 4 billion years. The earliest lifeforms that appeared on Earth constituted what we call prokaryotes. Prokaryotes are single-celled organisms that do not contain any structural components inside them such as nucleus or mitochondria. These structural elements inside the cell are called organelles, resembling little organs responsible for specific functions inside the cell. There were two types of prokaryotic species called bacteria and archaea. It has been concluded that in the early formative years of Earth around 4 billion years ago, the atmosphere of Earth was vastly different from its current state. It was much thinner with less density, and it contained very little oxygen if at all. Nitrogen, carbon dioxide, and water vapor were primary ingredients. As a result, these prokaryotes had to obtain energy using anaerobic processes by using chemicals available in the surrounding environment. Such prokaryotes are named chemoautotrophs. As such, chemoautotrophs did not need sunlight as well and could survive in harsh conditions. One of these bacteria or archaea was likely what scientists call *LUCA*, or Last Universal Common Ancestor, from whom all current life has originated.

With the appearance of these prokaryotes, Darwin's theory started to come into action. With each successive reproduction of these organisms, some variation of the original species started to appear and the offsprings started showing some new features and behaviors. With natural selection trimming the species that were ill-suited to the environmental conditions, the process kept on going for millions of years. About 500 million years into it, around 3.5 billion years ago, the first cyanobacteria appeared on Earth. They still belong to prokaryotes, but possessed an entirely new feature with them that enabled them to capture solar energy for their survival. Also, the process of capturing solar energy converted atmospheric carbon dioxide into pure oxygen as a by-product. As there was an abundance of carbon dioxide in the atmosphere, these species thrived and grew in epic proportions. Their growth had a profound impact on Earth. Their incessant consumption of carbon dioxide reduced its percentage from the atmosphere, reducing its greenhouse effect. As a result, the Earth started to cool down. It not only introduced oxygen into the atmosphere, but also increased its amount to the levels that enabled other species to use it for their energy production.

The next major step in evolution was the appearance of eukaryotes. Eukaryotes are also single-celled organisms, but they have a significantly more advanced cell structure that comprises organelles or structural components such as nucleus that contains DNA, mitochondria, or chloroplasts, etc. that have their own membranes. The evolution of eukaryotes from prokaryotes took about 1.3 billion years and is an extremely complex and fascinating topic that is still a part of active study among scientists. There are a number of different theories describing how eukaryotes could have evolved. The most widely accepted theory is called the endosymbiotic theory. The endosymbiotic theory suggests that eukaryotes evolved when multiple prokaryotic organisms physically merged into each other forming a single entity. It has been observed that some prokaryotes have a tendency to form symbiotic relationships with each other. In this symbiotic relationship, two organisms live close together and benefit from each other. In the case of eukaryotes, it is thought that one prokaryotic cell got engulfed into another prokaryotic cell. The engulfed cell was not digested or dissolved, but instead lived inside the host cell continuing to function and ultimately got converted into an organelle of the host species. There is a lot of evidence to support the endosymbiotic theory. For example, consider mitochondria and chloroplasts, two of the most common organelles found in eukaryotic cells. Not only are they similar in size and shape to other independent prokaryotic species, namely, alpha-proteobacteria and cyanobacteria, respectively, but also have their own DNA. This DNA is similar between them and is separate from the DNA in the nucleus of the host species. Even today we see many examples of plants and animals where such symbiotic relationship is observed between prokaryotes and eukaryotes. The bacteria provide their hosts with nutrients, and in return, they are protected from predators. The evolution of eukaryotes is a crucial landmark in the history of life on Earth. They exemplified a more complex cellular architecture and paved the way for the formation of multicellular organisms. Natural selection further allowed eukaryotes to become more successful than prokaryotes, and they eventually became the dominant form of life on Earth.

Multicellular Organisms

Based on the dating of fossils, eukaryotes are considered to have evolved and prospered around 2.7 to 2.1 billion years ago. The jump from eukaryotes to multicellular organisms is yet another important milestone in the evolution of species. However, definitive evidence of multicellular organisms has not been found until about 600 million years ago. This leaves a huge gap of anywhere from 2.1 billion years to 1.5 billion years, when the Earth was occupied by only single-celled eukaryotes and prokaryotes. However, it must be noted that absence of fossils does not make the appearance of multicellular organisms prior to 600 million years impossible. This is another area of active research and with every new discovery or new analysis of fossils found earlier is changing the currently established timeline of various species. So, these numbers are to be taken as rough estimates only. Even if these timelines can be approximate, the scientist community has a good understanding of how the transition to multicellular species must have happened. At the heart of this, is a same symbiotic relationship that led to the formation of eukaryotes from prokaryotes. The process is likely much simpler than what was thought earlier, and it is highly likely that it has happened not once but multiple times in the history of Earth. An example of Chlamydomonas is typically studied to explain this transition as it represents one of the latest transitions that happened just about 200 million years ago. Chlamydomonas is a simple eukaryotic organism that has a cell wall, a nucleus, and chloroplasts. It reproduces by dividing in half. As a first step in the evolution of Chlamydomonas, they started to form colonies. The colonies as such were still a group of many single-celled organisms. Over many thousand years, the different cells in the colonies started to take up different and specialized functions based on their location in the colony. Some cells became better at feeding the entire colony, while some other cells focused on reproduction and subsequent growth of the colony. With such specializations, the colony was no longer just a group of cells, but it now had a complex and well-defined structure. These specializations also helped the colony live longer and grow faster than the other colonies that were not going through these changes. Natural selection helped prioritize these colonies over time. After continued changes toward more and more specialized functions, cells lost their individuality as well as ability to survive on their own, ultimately leading to the formation of a true single multicellular organism called a sponge.

Arrival of Humans

After the birth of the first multicellular organism, the subsequent changes are much more streamlined and took place at a relatively breathtaking pace. To give a comparison, it took more than a billion years for the first single-celled prokaryotes to evolve into more sophisticated but still single-celled eukaryotes, while it took less than 400 million years for first multicellular species like sponge to evolve into first rat-like mammals known as Morganucodon. Mammals are an extremely complex and sophisticated species that exhibit advanced features like presence of hair or fur on the body for protection from the environment, being able to produce milk to feed their offsprings, along with advanced body structure comprising specialized jaws, backbone, and spine, along with an oxygen circulation system powered by a heart. Then the evolution of mammals like Morganucodon to primates in the form of Purgatorius or Archicebus took less than 150 million years. The primates mark yet another big milestone in the chain of evolution with even more advanced features such as forward-facing eyes giving binocular vision, hands with ability to grasp with fingers, larger brain, improved vocal tracts and communication abilities, and so on. The evolution from early primates to first hominin in the form of *Sahelanthropus tchadensis* took less than 60 million years; from first hominins to first of the Homo species (*Homo habilis*) took less than 5 million years; and from *Homo habilis* to *Homo sapiens* or modern humans took only 2.5 million years.

The evolution of the present generation of modern humans or *Homo sapiens* dates back to about 200,000 years with origins located in Africa. There is some debate among scientists around the precise timing and some believe that *Homo sapiens* appeared on the face of Earth as early as 300,000 years. It is well understood that the *Homo sapiens* co-existed as well as interbred with other related species such as Denisovans and Neanderthals for a certain duration of time, but ultimately eradicated the other species altogether. The differences between them were relatively minor, mainly pertaining to cognitive abilities. To give quantitative rationale, the similarities in the DNA between *Homo sapiens* and *Neanderthals* is about 99.7%, while between *Homo sapiens* and Denisovans is about 99.4%. To compare that with other species, *Homo sapiens* share 98% of their DNA with chimpanzees while only 97% with *Homo erectus*, making us genetically closer to chimpanzees than *Homo erectus*. In order for species to interbreed, the DNA needs to be extremely close, typically over 99%. Most modern humans carry about 2% DNA from Neanderthals, while people in Melanesia[2] inherit about 4% to 6% of their DNA from Denisovans.

Since then, humans have migrated all over the Earth and now have occupied almost all the habitable land across the entire planet. This spread is quite phenomenal and absolutely unprecedented in the history of life on Earth. No other species has been able to occupy the regions on Earth with so much diversity in climatic conditions, without spending hundreds of generations of evolution in those regions. Genetic scientists have found evolution of some minor but new traits being developed in humans living in different regions of Earth, but nothing close to the creation of a new species.

Over the last 100,000 years or so, the physical appearance of humans has not changed much. However, behavioral changes have certainly happened on multiple levels and at unprecedented speed, and they certainly don't follow the notion or rate of behavioral adaptation as seen in other animals. Let's consider an example to illustrate the differences.

If you pick a large enough group of lions (about a few hundred or so) from tropical forests and put them in the Antarctic region, most likely the whole group will perish in a matter of a few years if not less. In a similar manner, if you pick a similar group of polar bears from Antarctica and put them in Amazon forests, they will likely perish just like the lions in a matter of a few years if not less. The same can be said for most other animals, except for some microorganisms (as they are already adapted to survive in harsher conditions). Most larger species just cannot cope with the sudden changes in the environment or habitat of this magnitude and perish.

---

[2] Melanesia is a subregion of Oceania in the southwestern Pacific Ocean. It is made up of thousands of islands, including New Guinea, the Solomon Islands, Vanuatu, Fiji, and New Caledonia. The people in this region typically had dark skin, hence the name Melanesia comes from the Greek words melas, meaning "black," and nesos, meaning "island."

On the other hand, consider a situation where we pick a similar group of humans from Amazon forests and put them in the Arctic region, or vice versa. We also make sure that the group is not connected with the rest of the world in any way and cannot get help from there. Within their lifetime, the group of humans will start creating weather appropriate clothes, footwear, will build a new type of housing appropriate for the new climate, will create new weapons to hunt the animals in that region, will figure out plants and vegetables that they can use as food in that region, and ultimately will survive and even grow in population. They will likely not be happy at first, at least in the first generation, but nonetheless they will not perish.

Tools

This brings us to the concept of tools. This seemingly simple and casual example actually has rather far-reaching implications from the perspective of evolution. This example paves a way toward the fundamental shift away from Darwinian evolution. As a matter of fact, if humans *don't need to evolve* to adapt to the changing environment order of magnitude faster, there is no need left for their evolution. The main driving force toward evolution is adaptation and survival of the fittest species, but humans have created something entirely game-changing so that they can adapt and survive in any climatic condition that the planet has to offer while staying biologically and genetically *as is*. The secret sauce that helps humans achieve this incredible feat is the invention of tools. The very notion of tools is quite generic and vague, so let's bring some clarity to it. Tools can include anything that humans have created using their natural organs (e.g., hands, feet, brain, etc.) and leveraging living or non-living things available on Earth. Typical examples would include clothes, weapons, houses, automobiles, and even pets. The role of humans in creating pet dogs from wild wolves is well researched and documented.

The creation and use of tools marks probably the greatest achievement by humans, that puts them in a completely different league of species. No other species has shown the ability to create such tools with the agility that humans exhibit. There are examples of other species also using tools like chimpanzees using sticks to pick up insects, or using leaves to drink water from streams, etc. However, there has not been any changes in these tools over hundreds of thousands of years. Same can be said about birds that build nests, bees that build honeycombs, etc. However, the ability to create and/or use these tools by these species is attributed to evolution (that took hundreds of generations to materialize) and not adaptation (that can happen in a single to a handful of lifetimes).

Tools make humans independent from the need to alter their physical appearance. These tools augment human's core appearance and capabilities and adapt them to survive in the changed environment. The advanced brain of humans is also capable of changing behavior in a matter of days or even hours. We are capable of learning from past experience and can work in groups when needed or work alone when that is more effective, can communicate with each other to pass on the learnings in a matter of minutes and not generations, and so on. In other words, we can adapt in the same generation to such an advanced level that other animals need hundreds, if not thousands, of generations to achieve.

It should be noted that all the tools we have in the twenty-first century were not available over 300,000 years ago when humans first came into existence. It took a long time for us to get there. Although from the biological sense, humans were not quite evolving during that time, but the tools were.

Alternatives to Darwinian Evolution

So far, we have discussed the evolution of species from the perspective of Darwin's theory of evolution based on genetic variation and natural selection. Although this is one of the most widely accepted theories, there are some alternate theories that have been proposed. It would be interesting to investigate some of the top contenders among them.

If a theory is foolproof, there is no scope for scientific alternatives to it. For example, we can consider Newton's laws of motion. All the three laws have been proved beyond doubt by repeated real life experiments and observations of natural phenomena (unless we go to interstellar distances and speeds close to that of light) and as such there is no scope for any alternate theory to explain the physical interactions between objects. Even if Darwin's theory is so elegant and brutally simple, it is based on a sequence of logical arguments coupled with varied randomness over billions of years, unlike Newton's laws that are entirely based on mathematical equations with no scope for any randomness. As such it is not hard to poke holes in the theory of evolution. In spite of it being in such a delicate situation from the perspective of experimental proof, there have not been any fundamental flaws that have been discovered in it. Still there are some challenges that cannot be completely ignored. Here are some of the main criticisms: (1) inability to explain origin of life; (2) inadequacy in explaining inconsistent fossil records; (3) difficulty in explaining apparent teleology or evolution with purpose; and (4) difficulty in explaining adaptation vs. exaptation.

Darwin's theory really kicks in after life has already originated and as such the two underlying principles that drive the theory cannot explain how life came into existence in the first place. We looked at this in great detail in the earlier chapter, but none of the theories were based on evolution. This criticism is fair, but it simply points out the bounds on the phenomena that can be explained with evolution and does not really disprove it. However, one can claim that there is potential for a better and more general theory that can explain the origin of life as well as appearance of all the species on Earth in a streamlined fashion, much along the lines of string theory that tries to unite disparate physical laws such as quantum mechanics and relativity, etc.

Fossil records and their carbon dating are the only true real references or facts that we have to prove or disprove any theory. Unfortunately, the fossils that we have found so far are not uniformly distributed in time as well as locations, and there is a definite bias based on the availability of excavations in the regions. As per Darwin's theory, we should be able to see the gradual changes in all the features across all the species, but there are distinct gaps in fossil records that create missing links between different species and their evolutionary transitions. The existence of such gaps does not necessarily disprove Darwin's theory, but once again opens up a possibility for another theory to better explain these gaps.

Teleology

The concept of teleology assumes that the world as we see now has been created for a purpose. The theory of teleology states that all the organisms that ever existed as well as all the organisms that have gone extinct so far, did so for a purpose. It is more philosophical than scientific, and that certainly makes it a lot more controversial. The pursuit of finding the purpose in the creation of the world inadvertently takes us to the concept of God. The fact that humans are the dominant creatures on Earth, God must have created the world for humans to live and grow. As this theory does not state what is the ultimate purpose behind the creation of the world in terms of anything but what we see today, it is extremely hard to disprove. Any evidence that we may have obtained so far can fit with this theory. However, this theory has very little predictive power. In the case of any scientific theory, we should be able to change the parameters of the setup and the theory should be able to predict how the outcome would change. In the case of Darwin's evolution, if we assume that the catastrophic event causing the end of dinosaurs did not happen, the theory can predict what would the current Earth population look like. However, with theories like teleology, it does not matter what happened in the past to predict the present. It starts with the assumption that the present has to be what we see now! It uses a somewhat circular argument where the final result itself is used to prove how it came into being. One can prove any statement with a circular argument. Darwin's theory certainly cannot explain the purpose behind the current state of Earth other than pointing that it is a scientific and systematic outcome of two underlying principles combined with the chronological occurrence of astronomical events.

Exaptation

Exaptation is a word coined to explain the use of a trait or feature in a species that is evolved for one purpose but is being used for another one. A common example of exaptation is the use of feathers by birds. The feathers were originally evolved for providing warmth and insulation from the surroundings, but later they were used for helping them fly. Many such examples can be found with other species. This process draws a tangent to Darwin's theory of evolution. As per the theory, species evolve features that help them better adapt with the surroundings, but the inherent randomness in genetic properties can also lead to creation of features that are neither useful, not harmful, or the features that can have multiple uses.

Lamarckism

Now that we have looked at some of the top challenges or criticisms to Darwin's theory, let's look at some of the top alternatives that are proposed. One of the top alternate theories that has gotten some following is called Lamarckism, from its inventor, Jean-Baptist Lamarck. According to Lamarckism, not only the genetic traits are passed from parents to the offspring, but if the parents acquire some traits during their lifetime, they are also passed on to the offspring. To illustrate this concept, let's consider an example involving humans: As per Lamarckism, if both parents of a child spent a lot of time learning piano before birth of their child for many years and get really good at it, the child is likely to be born with fingers that are adapted to play better piano. This may sound quite intuitive, as historically we have seen that children follow their parents' business and over time, we see them getting better at it. Like a blacksmith's son becomes a better blacksmith, a carpenter's daughter becomes a better carpenter, and so on. However, the true reason the children get better in their traditional businesses is because they learn the tricks of the trade from their early childhood and get opportunities to improve on them. From a strict genetic standpoint, there has not been any proof that such traits can be transmitted.

Mutationism

Another theory is called mutationism. It borrows the concept of random genetic variation from Darwin's theory and takes it to the extreme. It states that mutations are not gradual but sudden. Big changes in species' features can happen in a single generation. If the changes are beneficial, they are passed on to the next generation, otherwise they are not. This theory can explain some of the gaps or inconsistencies in the fossils we have found so far, but in most other cases, the theory does not hold water. One of the major problems with mutationism is that it does not account for the gradual changes that we have seen in the fossil record, as well as the patterns of variation and adaptation that we observe in living organisms. Most mutations are harmful or neutral, and only a very small fraction is beneficial, which means that the chances of a large beneficial mutation arising are very low. Moreover, mutationism also fails to explain how complex adaptations, such as the eye or the wings of a bird, could arise through sudden, large-scale mutations. These complex structures are the result of a long process of gradual refinement and adaptation, rather than a sudden, miraculous change.

Intelligent Design

Then comes the theory of intelligent design. It claims that some of the features that we see in humans, or some other advanced species are too complex to appear as a result of pure randomness and natural selection. The theory tries to argue that statistically it is highly improbable for these features to have been created from scratch in about 4 billion or so years, hence there must be a supernatural being that is designing life on Earth explicitly. To actually work out the probabilistic math to find the probability of generation of humans from single-celled microorganisms is an extremely daunting task, but not impossible. If we consider all the intermediate species that we have found through fossil records, the chances of creation of humans are actually quite high. However, it is still based on a chain of uncertainties and one can easily argue against it. Ultimately this theory has proven to be less scientific and more philosophical.

Theory of intelligent design bears close similarity with teleology, but a key difference between them is the presence of the creator in intelligent design paradigm. In the case of teleology, such a creator is not needed.

Conclusion

We looked at the evolution of humans from single-celled organisms in this chapter, taking us forward in the journey to understand the nature and design of artificial intelligence, AI. In the next chapter, we continue this journey with exploration of humans got their intelligence.

CHAPTER 4

Human Intelligence

From the surging grasslands of Africa to the expansive stretch of our modern cities, the story of life is far broader in scope than just biological evolution. The adaptation of species by interacting with the environment and other species has led to an evolution in their body structure as well as the way they process information. As the Earth underwent countless transformations, the forces of nature meticulously sculpted our ancestors, guiding them from the primordial soup to bipedalism, from the basic instincts of survival to the intricate realms of introspection and creativity. This chapter delves into the origins, evolution, and profound depths of human intelligence. We explore the intricate details of the events and adaptations that have made us the thinking, imaginative beings we are today. Understanding human intelligence would pave the way in understanding how we can emulate it through the use of computing machines toward the development of artificial intelligence.

What Is Human Intelligence?

Humans are definitely the most intelligent and superior species on the Earth as we have proved by our total dominance, but they are not the only intelligent species. All the living organisms all the way down to single-celled ones show some definitive signs of intelligence. They may not be able to converse or explain what they are doing and why they are doing, but nonetheless they are exercising rather sophisticated biological

© Ameet Joshi 2024
A. Joshi, *Artificial Intelligence and Human Evolution*,
https://doi.org/10.1007/978-1-4842-9807-7_4

processes through the way of homeostasis and staying alive and thriving in hostile environments; and this is intelligence! The overarching definition of intelligence is multifaceted and is a hotbed for debate and ever-growing theories. However, the most widely accepted features that define "humanlike intelligence" include the ability to learn from experience, adapt to new situations, understand and handle abstract concepts, and use knowledge to manipulate the environment. However, the notion of abstract concepts is tied quite intimately with the human brain, and we cannot relate it meaningfully to other animals. Hence, we need to expand the concept of intelligence that can apply to all the species on Earth for doing apples-to-apples comparison. With appropriate corrections, the defining features for intelligence reduce to the ability to perceive the environment, process information, and respond in ways that maximize chances of success or survival.

As per this broader definition, intelligence is not necessarily tied with the processing power of the brain, as there are thousands of species that can survive and even thrive on the Earth that do not even possess the brain as a separate organ. It is also not tied with brute physical strength, as there are so many animals that are significantly bigger than humans but cannot compete with humans when it comes to survival. The example of woolly mammoths, which were 100 times bigger than humans is pertinent here, as humans played a major role in their extinction. So, what is it then that makes the difference? What is it that makes one species more intelligent than another?

The earlier definitions of intelligence focus on the question of "what" or the effect of intelligence and do not shed much light on the "how" or the causes of intelligence. So, let's take a look at intelligence from the other. From this perspective, intelligence is defined as "the ability to acquire and apply knowledge and skills." The ultimate goal of this process still remains as maximizing the chances of success and survival, but now we can look at how different species achieve that. The words knowledge and skills are still quite broad and do not necessarily convey an obvious and clear message. However, pursuing this definition can certainly help unravel the mystery of intelligence better than the earlier definitions. There are two distinct sets of concepts involved here: acquisition of knowledge and application of it in the form of display of skill. Acquisition of knowledge refers to various sensory organs and the way they acquire information about the surroundings. Application of knowledge is a process of using this information to solve problems, make decisions, and to better react to the environment. Both acquisition and application of skills are part of the later process, and we will deal with them together.

Acquisition of Knowledge

In the next sections of this chapter, we will look at the sensory system of humans and compare and contrast it with other organisms with the intention of identifying if we can find something that would distinguish humans from all the other organisms that make us superior. Humans have five primary senses: hearing (audition), sight (vision), taste (gustation), smell (olfaction), and touch (tactility). Most other species share a similar if not identical set of sensory organs. However, there exist some species that have some different and or an entirely new set of sensory organs. Based on this feature, one can state that species with higher intelligence can acquire more, or better knowledge and skills compared to other species. Let's see if humans possess an obvious edge with respect to acquiring knowledge from the environment in a better way.

Hearing (Audition)

Human hearing is typically considered to be bounded by a range of sound frequencies from 20 cycles per second to 20,000 cycles per second. A cycle is a sequence of compression and rarefaction of air molecules as shown in the Figure 4-1.

Figure 4-1. *Sequence of compression and rarefaction in a sound wave*

Cycle per second is also called Hertz or Hz. Hence 20 Hz means 20 such cycles are produced per second and 20 kHz means 20,000 such cycles are produced per second. Thus, a human ear can sense such sound waves as long as their frequencies fall in the audible range. Human audible range starts deteriorating with age and a realistic range for a 30-year-old adult drops from 20 Hz to only about 15 kHz. The maximum auditory range is reached at about 8 years of age. The range of frequencies that we can generate with our vocal tract is actually a very small portion of the audible range and spans roughly from 50 Hz to 3 KHz. Thus, even with reduced auditory range, humans can perceive all the verbal communication with other humans. Most sounds that we encounter in day-to-day life that matter to us such as sounds of musical instruments, animal sounds, sounds from nature such as the blowing of the wind, rain, waves from the ocean, or machine sounds such as cars driving, trains rumbling, airplanes flying, and so on, all fall well within the reduced audible range and humans of all ages can perceive them.

Many other species also possess the ability to hear sound waves that match and even in many cases surpass human ability by a long shot. For example, dogs can sense audio frequencies from 40 Hz to 60 kHz, cats can hear from about 50 Hz to 85 kHz, elephants can hear from all the way down to 1 Hz to humanlike 20 kHz. The sounds with frequencies more than 20 kHz are collectively called ultrasound. Bats with ability to sense from 1 kHz all the way to 200 kHz, can use ultrasound to echolocate the placement of their surroundings based on complex patterns of the reflected sound waves that they generate with their mouth. This represents almost an entirely new type of sensory capability even if similar sound waves are used. Even aquatic organisms such as dolphins can sense sound waves from 75 Hz to 150 kHz, while whales can sense from 10 Hz to 100 KHz. To put simply, humans do not come out with any special ability in this category at all, rather they possess quite a mediocre part of the whole range of sounds that can be perceived by other organisms.

Sight (Vision)

Similar to hearing, human vision is also limited to a range of frequencies that begins from a staggering 400,000,000,000,000 Hz or 400 TeraHz or 400 THz to 750,000,000,000,000 Hz or 750 THz. It is important to remember that these frequencies are in the electromagnetic realm and not directly comparable to frequencies of sounds. Electromagnetic waves are composed of alternating crests and troughs of electric and magnetic fields perpendicular to each other as shown in the Figure 4-2, and they do not need any medium for travelling, such as sound waves.

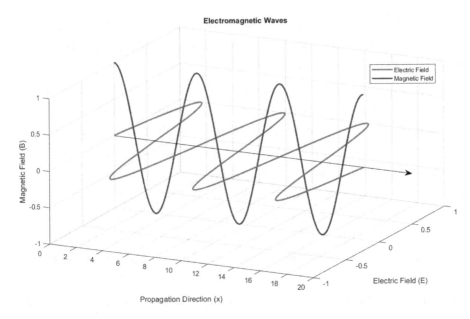

Figure 4-2. *Electromagnetic wave showing electric and magnetic waves perpendicular to each other*

The frequencies near 400 THz are perceived as red color by human eye, while the frequencies near 750 THz are perceived as violet color. All the intermediate frequencies create a color pallet that is broadly classified into seven colors such as red, orange, yellow, green, blue, indigo, and violet. Frequencies lower than 400 THz are called infrared frequencies and higher than 750 THZ are called ultraviolet, and both these types are invisible to the human eye. Even though our eyes are completely blind to infrared waves, they are perceived as heat by the human body through the sense of touch, but we don't have any directional information as to from where the heat originates the way we get for visible light. There are some other species that possess the capability to sense the light that falls into infrared as a visible signal. For example, frogs or snakes can sense the infrared rays from the surroundings and it is a critical ability that enables them to hunt their prey. Eagles have a highly developed vision system as well and they can sense even ultraviolet light through their eyes. Pigeons also possess a highly advanced vision system with a visible range spanning from red to ultraviolet. They also have better motion detection capability with faster response to changing lights as well. They may also be able to see magnetic fields that help them navigate longer distances, but this aspect of their capabilities is not fully understood yet. Even from aquatic species, there are certain types of shrimps, called mantis shrimps that can detect 10 times more colors than human eye, including ultraviolet light.

With respect to the capabilities of vision, the range of spectrum only paints one side of the picture, and there are many other facets that need to be looked at. For example, the clarity or resolution of vision. This feature refers to how our eyes can discern minute objects when we look at them. In this department, eagles have about 8 times greater resolution than humans, enabling them to spot their prey from several kilometers away. Then there is also the ability to see in low light situations. Cats and dogs, along with pigeons, have advanced night vision capabilities, enabling them to see with high resolution in very low light situations where humans are nearly blind.

Then there is another aspect: the number of eyes and their placement. Most animals and birds and even aquatic organisms possess two eyes, but there are some exceptions such as spiders who have eight eyes or horseshoe crabs that have ten eyes or box jellyfishes that have an astounding 24 eyes. Among the species with two eyes, the placement of the eyes also makes a crucial difference in the way surrounding image is perceived. One type of placement is where the field of view of each eye does not overlap with the other and eyes are spaced far apart from each other pointing in different directions. This vision is called monocular vision. With this type of setup, the organisms generally get a much wider view of their surroundings without moving their head. Examples of species with monocular vision are rabbits, deer, horses, frogs, most fish, and birds. It is important to note that the images formed by the eyes are two dimensional, while the world around us is three dimensional. Hence just by looking at one image one cannot discern the relative distances of the objects in the view. With monocular vision, as the two eyes are seeing entirely different things, the species with this type of vision lack the understanding of depth. The other type of setup is called binocular vision or stereo vision. In this setup, the two eyes are placed relatively close to each other, pointing in a forward direction. With stereo vision, the views from both the eyes overlap. The drawback of this setup is the reduction in the width of all the surroundings that are captured in one position. Hence to capture a wider view, such species would have to rotate their head. However, there is one big advantage with this setup. When two eyes see the same objects from slightly different angles, based on the subtle differences in their shapes and shadows perceived by each eye, the brain can process this information to generate the depth information about the surroundings. This process is called stereopsis. The depth information calculated by the brain is still an estimate and can have some errors in it. The better the surroundings are lit, the better the depth estimation. This information about depth is critical for the organism if it belongs to the category of predators. All the predatory organisms on planet Earth have stereo or binocular vision, while all the organisms that belong to the prey category possess monocular vision. Humans belong in the predatory category with roughly 75% of the horizontal overlap between the field of view from both eyes. This strong overlap allows humans to perceive the depth of field quite accurately. In general, stereo vision is one of the most advanced vision setups observed among all types of species. Humans do belong in this category, but again, they are nowhere near the top. Birds such as eagles, hawks, and owls or even land animals such as dogs and cats possess heightened capabilities of stereo vision in one of the other aspects.

Taste (Gustation)

The taste or gustatory system works in a different way compared to other sensory systems that we looked at. When the molecules from food or any other object come in direct contact with the gustatory sensors or taste buds, a chemical reaction takes place that generates certain neural signals that are then transmitted to the brain, which is construed as taste. The human taste or gustatory system, has evolved to detect a variety of tastes, which historically served as mechanisms for survival by helping differentiate between safe and potentially harmful substances. Humans have about 2000 to 10,000 taste buds, primarily located on the tongue, but also on the roof of the mouth and in the throat. Each taste bud contains anywhere from 50 to 100 taste receptor cells. Equipped with these, humans can taste five primary taste modalities in the form of sweet, salty, sour, bitter, and savory. Most organisms on land, water, as well as birds possess some form of taste buds that serve a similar purpose. Perception of taste is quite subjective, and one can argue that some animals may have less or more than humans, but from the objective comparison of number of taste buds, there are numerous examples having significantly more taste buds than humans. Catfish have an extraordinarily large number of taste buds, upwards of 100,000 are distributed over their entire bodies, while cows have about 25,000 taste buds, rabbits have 17,000. Some other organisms such as fruit flies have a sophisticated gustatory system with receptor neurons distributed in their legs along with mouthparts, giving them the ability to taste outside of mouth. Once again, humans don't come out as outliers or dominant species from the perspective of acquisition of taste.

Smell (Olfaction)

Smell or the olfactory system enables an organism to perceive yet another type of information from the surroundings that is completely orthogonal to the information acquired from auditory, vision, and taste, although it bears some resemblance and correlation with taste. The smells originate through the chemical compositions of the materials present in the surroundings. Some molecules from these chemicals evaporate at room temperatures and travel through the air. When they come in contact with the olfactory sensors, a sense of smell is created. The human olfactory system, responsible for our sense of smell, plays a crucial role in numerous aspects of our daily lives, from food selection to emotional experiences.

Measurement of the capability of detecting smell is quite complex compared to other modalities. One can always count the number of olfactory receptors, but that does not paint the full picture. Another metric is used sometimes to measure olfactory capabilities in the form of lowest concentration of a particular substance that an organism can detect. This is called the detection threshold. The lower the threshold, the more sensitive the olfactory system. Another method tests the ability of an organism to distinguish between different odors. Along with the ability to distinguish odors, remembering the odors is also another dimension in measuring the olfactory capability. Comparing humans with other organisms based on these parameters, humans have about 5–6 million olfactory receptors and based on genetic analysis, we can identify about 400 different odors. However, dogs have a staggering 300 million olfactory receptors along with about 1000 olfactory genes, giving them the ability to detect a vastly broader spectrum of odors with order of magnitude higher sensitivity. Similarly, elephants have over 2000 olfactory genes, rats have over 1500 olfactory genes, giving them enhanced olfactory abilities. Thus humans again come out with mediocre olfactory capabilities.

Touch (Tactility)

Last but not least, touch or tactile system is our fifth sensory system and one of the simplest to understand. The receptors for touch are present all over the skin of the organisms. While it may appear simple superficially, its operation is rather complex and encompasses a variety of receptors responding to different types of stimuli. These receptors transform mechanical energy, temperature, pressure, and various other stimuli into electrical signals and send it to the brain. The primary tactile receptors include (1) mechanoreceptors, capable of detecting pressure, vibration, and stretch. These include Merkel cells that respond to sustained pressure and texture, Meissner's corpuscles that detect light touch and grip control, Pacinian corpuscles that sense deep pressure and vibration, and Ruffini endings that detect stretch and twisting. (2) Thermoreceptors, capable of detecting changes in temperature. These include Warm and Cold receptors. (3) Nociceptors, capable of detecting pain, whether from excessive pressure, temperature, or chemicals. And (4) Proprioceptors, which are located in muscles, tendons, and joints and they provide a sense of body position and movement.

While humans possess a refined sense of touch suitable for intricate tasks and the perception of their environment, many animals have evolved specialized tactile systems adapted for their specific ecological niches. The star-nosed mole's ultra-sensitive nose, the cat's whiskers, or the lateral line system in fish all exemplify the diverse adaptations of the tactile sensory system in the animal kingdom.

Other Sensory Systems

Furthermore, there are examples where some species have an expanded range of sensory organs that are entirely missing in humans along with many organisms with similar sensory system, but with an order of magnitude higher capability. For example, dogs can smell nearly 20 times better than humans making them best detectors of drugs or an invaluable addition in the hunt of an individual or animal. This ability almost makes their olfactory sense as an entirely new type of sensory channel compared to humans. Sharks and some other fish possess specialized organs called "ampullae of Lorenzini" that can detect minute electrical fields produced by the movements or the bioelectric fields of prey. This aids in hunting, especially in murky waters. Some birds can sense changes in the Earth's magnetic field, helping them better navigate during continental migrations and so on.

Thus, looking at the overarching sensory systems spread across all the organisms, humans fall somewhere in the middle, with nothing really extraordinary. However, one can still argue that even if humans do not possess a clear advantage in acquiring knowledge in one sensory area, as a combination of all areas or modalities humans may have an advantage. This still remains a highly subjective aspect and as such not quite conclusive.

Application of Knowledge and Skills

As we saw in the earlier section, there exists a diverse spread of sensory capabilities across different organisms that are adapted to survive in their respective environments. Humans do not show any particular advantage in either of the categories. Let us see in this section how humans process this information to generate skillful actions and compare and contrast it with other organisms to see if we observe any indicators for human superiority. Although the sensory organs are spread all over the body, the processing of sensory information mostly[1] happens in the brain for humans as well as other species. Do humans have an extraordinarily large brain then? Or an extraordinary number of neurons crammed into a small size? This has been the area of active research for a long time. A typical human brain weighs about 1.2 to 1.4 Kilograms and contains about 86 billion neurons. In comparison, the elephant brain weighs about 5 kilograms and possesses about 257 billion neurons. So, on the brute scale of the number of neurons, humans don't come on the top. Looking at a closer relative, a typical chimpanzee brain weighs a measly 350 to 450 grams and contains only 28 billion neurons. How about looking at another

---

[1] Although most responses generated by humans and other organisms are generated in the brain, there is a whole other set of responses called involuntary responses that are generated by autonomous nervous system. They include reflex actions, goosebumps, salivation, and so on. These actions are critical and can help survive sudden and sometimes life-threatening situations.

measure in the form of the number neurons per unit body weight? This measure is commonly called Encephalization Quotient, or just EQ (not to be confused with emotional quotient). The EQ for humans is quite high compared to many animals and primates at 7.4 to 7.8. In comparison, most mammals have EQ of around 1. Bottlenose dolphins, who are also highly regarded for their intelligence, exhibit an EQ of around 6.3–6.7, while pilot whales also possess an EQ of over 6.0. Overall, EQ metric shows some clear advantages for humans, but it is not considered as a decisive and trustworthy feature. Some small animals like birds and rodents can show higher EQ numbers (not explicitly proven yet), but that does not make them intelligent. Along with the sheer size of the brain, its composition and density and complexity also play crucial roles in determining the intellectual prowess of the organism.

Thus, there is no conclusive metric that puts humans at the top of the pyramid, and once again we need more investigation. Let us look at how humans can use their average sensory system with the help of an average sized brain and still exhibit extraordinary features and skills.

Extended Fluidity in Human Brain

Even if the human brain is not extraordinarily big or contains the highest number of neurons, its operation is not exactly similar to many other species. Humans are genetically predisposed to having more fluidity in the development of their brains. In technical terms, the human brain possesses higher neuroplasticity than most other organisms, meaning the human brain can change and adapt its structure and function in response to experience more so than any other organism. As a result of this ability, humans can learn new things and adapt to new situations throughout life, while such adaptation is severely muted in other organisms. Added to this, the relatively large size of the human brain and its complexity enables humans to be in an order of magnitude more adaptable compared to other species. The longer lifespan of humans furthers this adaptation.

In practical terms, in humans, a significantly smaller portion of the brain structure is inherited from the parents compared to other species. Each human is given a gift of building their own brain structure that can deviate significantly from that of their parents. Other species are not privy to this and in their case, the structure and resulting functioning of their brains is pretty much set with their DNA and genetic makeup. Only a minuscule number of changes can happen through mutation and can create new features only at birth, but they are extremely small compared to what is possible with the human brain in a single generation. Consequently, a significant portion of the development of the human brain happens after birth, while in most other species, the development of the brain is nearly complete at the time of birth. Due to this, the environment in which a human child is raised plays a much bigger role in shaping the brain of a human child compared to the children of other species. One drawback of this feature is that human babies are significantly more vulnerable to the surrounding environment during this process than babies of other species, as human babies don't even possess some of the basic survival skills at the time of birth that other babies are born with. However, the advantage of this feature is that a human child can be significantly better and different than their parents in adapting to the given environment. This creates an unprecedented ability in human babies to learn new things that their parents are not even aware of. All the other species on Earth lack this feature almost entirely. Thus, humans start to emerge as a superior species with their extended adaptation capabilities along with a rather compelling mix of biological and sensory capabilities when everything is put together.

Skills

Now, let's take a look at the acquisition and application of skills. "Skill" as such is a generic term that tries to capture all the aspects of accomplishing a task. In order to give more color to the concept, let's consider some examples of skills: a skill can be to climb a tree, or to be able to run at fast speeds or to be able to chew and digest meat or to be able to pluck a flower without breaking the branches around it and so on. The skills can range from being too ordinary and superficially frivolous to being critical in survival. When a species is particularly good at the latter type of skills, they become stronger candidates for being able to live longer and prosper. However, in most cases, all the skills that a species possesses are required for their well-being one way or another. The sensory abilities that we discussed earlier in the chapter help in carrying out the skillful jobs better. To compare humans with other species, let's first identify the key survival skills. For any species to be able to survive, the first and foremost skill that is needed is the ability to find and consume the ingredients from the environment that provide energy. Typically, on the Earth, this means to be able to breathe oxygen from the atmosphere, be able to find, eat, and digest food and drink water. The second most important skill is to be able to reproduce and be able to nurture the offspring till it becomes self-sufficient and capable of carrying out the first skill. Any species equipped with these two skills should be able to survive and grow in their respective environment given they are the only ones living there. However, when we look at the full picture, there are many other species also living in the same environment, competing for the natural resources or are dependent on other species for resources. In such a case, there is one additional skill that is required and that is the ability to protect themselves from other species. If a species is not able to protect itself from another competing species or from a predatory species that may use it as food, it is likely to perish in a short time. Equipped with these three basic skills, any species can have a fighting chance to survive and grow. However, pretty much all the species that we see on the Earth possess all the three skills, and humans cannot claim to have an edge with respect to them. Any species that start to become weaker in either one of the skills due to change in the environment or due to change in the species around them, would likely become extinct within a few generations. There are millions of species on the Earth and out of them thousands are facing extinction due to such changes around them. On top of that, there are thousands of species that have already gone extinct during the past few hundred years that we have kept records of. It will be interesting to look at some examples of the species that have gone extinct in the recent past.

Recent Extinctions

Dodos

Mauritius is a beautiful island nation in the Indian ocean, east of Madagascar. It is known for its pristine beaches, crystal clear waters, and diverse wildlife. The dodo birds lived on Mauritius for thousands of years. They were unique and fascinating creatures, and their extinction is a very sad loss. Extinction of dodos was fueled by a combination of factors, such as human hunting, habitat loss, and unnatural introduction of predatory species. The dodo was a slow, flightless bird that was easy to kill. Humans hunted the dodo for food and sport. On top of that, the dodo's habitat was destroyed by deforestation and the introduction of new species from the humans. The dodo needed a specific type of forest to survive, and when this forest was destroyed, the dodo had no place to live. To make matters worse, the dodo was also not able to compete with the invasive species such as rats, dogs, cattle, pigs, and monkeys that were introduced to Mauritius by humans in a quite unnatural and drastic manner. With the addition of these species, it became difficult for dodos to find the food they needed, impacting the first necessary skill. Then human hunting impacted their third skill to protect themselves. When changes to the environment happen at a natural rate, the species generally get sufficient time to adapt/evolve to the changes. This is the basic premise of evolution by natural selection that we looked at in the earlier chapter. However, humans played a key role here, changing things too fast and not giving dodos enough time to evolve. As a result, they became completely extinct by the end of the seventeenth century.

Cisco Fish

Now, let's look at another example in a completely different habitat, deep waters. Cisco fish is a type of salmonid fish that is found in the Great Lakes and other bodies of water in North America. They are a popular "sport fish."[2] Cisco fish are known for their silvery color and their long, slender bodies. They are related to other salmonid fish, such as salmon and trout. Cisco fish are an important part of the Great Lakes ecosystem. They used to be a food source for many other animals, including birds, fish, and mammals. They were also valuable commercial fish, and they were used to make a variety of products, including fish oil, fish meal, and smoked fish. However, some varieties of cisco fish are not entirely extinct. In spite of the different environment, the primary reasons for their extinction are quite similar to the ones responsible for the extinction of dodos, such as overfishing by humans, presence of invasive species, and loss of habitat. Due to high commercial value, these fish were captured at a rate that was not sustainable based on their reproduction cycle and hence their population started to drop significantly. Then, to make matters worse, a new breed of invasive species was brought into the same habitat. Here humans did not bring them on purpose as they did with dodos, but it was more of an indirect effect. A canal called "Welland Canal" was built in the early 1800s to connect Lake Ontario and Lake Erie. The canal was built to allow ships to avoid Niagara Falls, without which the ships could not pass from one lake to another and carry the big cargo across, creating a critical bottleneck in transportation. The first version of the Welland Canal was built in 1829 and it has been expanded multiple times since then. Before the existence of this canal, most of the lakes in the Great Lakes system were isolated from each other and especially from the Atlantic Ocean,

---

[2] Sport fish is a fish that is caught as a sporting activity rather than for the purpose of eating. These types of fish are not easy to catch, and they offer strong resistance to capture. So, catching one becomes a challenge and reward. Typically, they are also beautiful and rare.

except for Lake Ontario, which was connected with the Atlantic Ocean. With the advent of the Welland Canal, the Atlantic Ocean got directly connected to all the lakes and, as a side effect of this, a brand-new breed of an invasive species in the form of seal lamprey was introduced into the lakes' ecosystem. These species made the life of cisco fish even harder, which was seeing dwindling population due to overfishing. On top of this, the canal also impacted the habitat inside the lake waters, which was made worse with deforestation, creation of dams, and pollution. As a result, some species of the cisco fish became extinct.

Early Extinctions

Humans have played a leading role in the extinction of many species over the last several decades. However, humans are not the only cause of extinction that has occurred repeatedly through the billions of years of history of the Earth. There are a total of about five major events where thousands or even millions of species on the Earth became extinct. These events are called events of mass extinction and are named based on the geological eras that they separate. Here is the list of these mass extinctions: Ordovician-Silurian, Devonian, Permian-Triassic, Triassic-Jurassic, and Cretaceous-Paleogene. These extinctions are identified based on the age of fossils found across the Earth. Around the timeline of all these extinctions, we can clearly observe that the majority of species went extinct in a very short period of time.

The Ordovician-Silurian mass extinction occurred about 445 million years ago. It is the first of the five major mass extinctions in the Earth's history, and also the second largest extinction event known to us. This extinction event wiped out an estimated 85% of all marine species and about 70% of all terrestrial species. The cause of the Ordovician-Silurian extinction is not fully understood, but the most widely accepted theory puts glaciation and global cooling at the root. The ice age associated with this event cooled down the Earth's temperature significantly, increasing the polar ice caps and reducing the water levels across the globe. As a result of, many marine and especially coastal ecosystems were entirely wiped out. Changes in the atmosphere are thought to have caused a decrease in oxygen levels, which would have made it difficult for many terrestrial animals to survive. Changes in the ocean are thought to have caused a decrease in salinity, making it difficult for many marine animals to survive. The Ordovician-Silurian extinction had a profound impact on the evolution of life on Earth. Many groups of animals that were wiped out during the extinction event never recovered, and all the groups that survived went through significant changes through fast paced evolution. The extinction event also led to the rise of new groups of animals, such as the modern-day fish and the amphibians.

The Devonian mass extinction was the second of the five major mass extinctions and occurred about 375 million years ago, during the Devonian period. The extinction event wiped out an estimated 70–80% of all marine species and 20% of all families of Devonian animals. The cause of the Devonian mass extinction is also not fully understood. An asteroid impact or large volcanic eruption is considered the primary root cause. However, either of them manifested into a similar combination of factors, including a sudden drop in temperatures, changes in the atmosphere, changes in the ocean that ultimately led to mass extinction.

The Permian-Triassic extinction event was the third mass extinction that wiped out more than 90% of all marine species and 70% of all terrestrial vertebrate species. It is by far the most severe known extinction event, and it occurred about 252 million years ago. Unlike the previous two extinctions, there is a consensus among scientists that the root cause of this extinction was a large increase in atmospheric carbon dioxide due to massive volcanic eruptions in the region of Siberian Traps in present-day Russia.

The fourth mass extinction or the Triassic-Jurassic extinction was also triggered by widespread volcanic eruptions around 200 million years ago and resulted in similar conditions with elevated carbon dioxide, global warming, and killed off over 75% of marine and terrestrial species.

The final and most recent mass extinction occurred about 66 million years ago and is responsible for the eradication of dinosaurs. The root cause of this mass extinction is nearly confirmed as an asteroid impact. Not only do we know the reason but we also have identified the place of impact as the under-ocean crater in the gulf of present-day Mexico's Yucatan peninsula named as Chicxulub crater. The asteroid impact caused a massive explosion that sent shockwaves through the Earth. The shockwaves caused a massive tsunami that devastated the coasts of Mexico and the Caribbean. The impact also covered the entire Earth's atmosphere with dust and debris from the impact explosion and blocked out the sun for months. Most of the species could not survive these harsh conditions for months and became extinct.

The mass extinctions clear out almost an entire living population on the Earth and reset the process of evolution and create a new starting point for the creation of whole new species. However, the role that humans are playing in selective extinction of species over the past decades, centuries, and even millennia is quite unprecedented, and it shows that somehow humans have become so far more advanced than all the other species on the Earth combined and that is just unfair.

So, given that humans have the three basic skills just like most other species on the Earth: ability to find food/water, ability to reproduce, and ability to protect themselves as well as the sensory abilities similar to most species on the Earth, our superior brain is capable of something special that other species just cannot compete with. In the next part of this chapter, we will investigate this secret sauce created by humans.

Tools

Most species other than humans use their natural body to perform the skilled operations required for their survival; however, humans have started augmenting their natural body with tools to expand the scope of operations they can perform. This might seem like a minor difference, but over time, it has become the primary distinguishing factor between humans and all the other species. Over the last several thousand years, rather than adapting their natural body potentially through the process of evolution, humans have been adapting and evolving their tools!

Mention of tool makes one think about the artifacts that are made from rocks or metals such as an axe or a sword, but they need not be this advanced. A basic tool can be as simple as a stick taken from a broken tree, or a stone picked from the ground, or a leaf cut from a tree and so on. The purpose for which an object from the environment is used makes that object a tool. This object now marks an extension of the human body that helps them to do something that is not possible with the bare human body. A stick can help a person to climb farther on a hill, the act of throwing the stone picked from the ground can help a person scare away a bird, a leaf cut from a tree can help the person cover up a scratch on their foot, etc.

Use of basic tools such as sticks and leaves is not strictly limited to humans, but many other species have been observed to use some form of tools. The most basic form of tools are the artifacts that are readily available such as thorns or conveniently shaped branches. Many birds are known to use these naturally occurring tools to capture prey or to help with consuming prey. A more advanced way of using tools involves custom modifications or assembly of the naturally occurring objects to achieve a specific goal. For example, many birds use small twigs and leaves to make nests to lay eggs. Nests are a definitive form of an advanced tool that helps better protect the eggs from the environment as well as other predators. Many primates have been observed to make extensive use of tools. Chimpanzees are known to use sticks to "fish" for insects such as termites and ants. They also have been known to use stones to crack open nuts and seeds. Orangutans use sticks to reach fruits that are out of reach. Chimpanzees also use sticks and stones to fight with each other as well as to defend themselves from other predators.

Use of tools is not limited to land animals and birds, but marine species also exhibit rather innovative use of tools. Some varieties of orcas are known to use dead fish as bait to attract other fishes, thereby using the dead fishes as tools. Octopuses are known to use rocks and sticks to open clams as well as to escape from predators.

In spite of this broad definition for use of tools by other species, there is a key difference in the way these tools are used. If we look back in history, use of tools by all these species has not changed over hundreds or thousands of years. There is definitely some adaptation based on changing habitats, but in principle, the use of tools by all other species is limited to finding naturally occurring artifacts and using them as is. We may see that birds living in cities may use some leftover debris from building construction that looks like a tree branch, but there is little evidence that a bird has fundamentally changed the structure of its nest to better adapt to changing habitat. We never see chimpanzees start improving on the design of sticks and explicitly make them sharp or that use them with a bow to hunt other species at a distance, even if they have arms that are more than capable of doing so. The early humans started with the use of simple tools like sticks and stones just the way we see other primates use. The oldest fossils of primates[3] date back to over 55 million years, much before that of humans. However, in spite of the chronological advantage, they still

---

[3] Primates represent a specific type of mammals that includes apes, monkeys, and prosimians (this group contains species such as lemurs, lorises, and tarsiers) along with humans. They are characterized by a set of evolutionary traits that reflect adaptations to life in trees, although not all primates are arboreal in the modern era. The order of Primates is diverse and includes the following general characteristics and groups: opposable thumbs, forward facing eyes, large brains, flexible limbs, extended parental care.

use the same basic tools, while in a relatively short time (still hundreds of thousands of years), humans were able to build on top of this and incrementally improve these tools to make them more effective.

In general, the level of modifications that most species can do on top of what is naturally available is quite minimal and mostly random. What this means is that the use of tools by all these species has been part of **evolution** and **not adaptation** and is etched into their DNA that is passed from generation to generation. Instead of parent birds teaching their children how to build a nest, or the bird experimenting with different options, the skill of building a nest is passed on from generation to generation through genes that have evolved over thousands or hundreds of thousands of years. This pace is quite slow compared to the pace of adaptation. This is where humans start to show marked differences, and the extended fluidity of the human brain after birth as described earlier plays a key role. Rather than waiting for evolution to show its effects, humans can adapt to the changed situation in as fast as a single generation by just improving their tools. With unprecedented speed in improving tools, humans have come up as a strong and dominating species that has become so advanced and powerful over time that other species just cannot match. However, humans too did not build the self-driving car or the Burj Khalifa Tower in one day or one year, it took tens of thousands of years for humans to reach the level sophistication. But progress was still order of magnitude faster than what evolution can depict and, more interestingly, it's not just the tool technology that is improving, but the rate of its improvement is increasing as well.

The notion of the word humans, here, is not necessarily meant to imply the modern humans who are named as *Homo sapiens*, but all the earlier members of the *homo* family that are extinct now. It started with *Homo habilis*, who lived in Africa around 2.6 to 1.8 million years ago. *Homo habilis* evolved into *Homo erectus*. This is the first species in the *homo* family that started migrating to other regions such as present-day Europe and Asia. They lived between 1.8 million to 117,000 years ago. Then there are multiple species such as *Homo neanderthalensis* or what we call the Neanderthals who lived in Europe and northwestern Asia around 400,000 to as recent as 40,000 years ago, *Homo floresiensis*, who were only found on the small island of Flores in Indonesia around 100,000 to 50,000 years ago. There are also many other lesser-known *homo* species identified by scientists, but their existence is still under debate. Sometimes the species such as *Sahelanthropus tchadensis* (between 7 to 6 million years ago in Africa), *Orrorin tugenensis* (around 6 million years ago in Africa), etc. going up to *Australopithecus afarensis* (around 4 to 3 million years ago in Africa) are also grouped under the *homo* family. *Australopithecus afarensis* have a special place in the whole list as they are thought to be the closest relative to *Homo habilis* and might have even interbred with them. They were fully bipedal and omnivorous and lived in a variety of habitats. One of the oldest and most complete fossils of *Australopithecus afarensis* was found in Ethiopia called Lucy. Lucy lived about 3.2 million years ago and was 3 feet 6 inches tall. The notion of *humans* in this chapter refers to this entire family of *homo* species.

Consider a special day some three million years ago, when a group of humans was hunting a pack of wild goats, with their regular blunt sticks as they used over the years, the stick from one member, let's call her Maya, broke as it hit a tree. The broken piece was a bit shorter but had a significantly sharper edge than the original one. When Maya tried to check the broken part of the stick with her own hand, she felt a sharp pain with a cut. She suddenly realized that the broken piece had become an even better piece of weapon. She showed it to all the other members in the group. They all tried to break their sticks to create the more potent weapon. The hunting that followed that day was unprecedented. They were able to hunt many goats with much less effort. From that day, that particular group, led by Maya's accidental discovery, advanced to the next level in hunting! They all updated their sticks for hunting, improving their hunting abilities. However, keeping the edges of stick sharp was hard and with little use they would lose their sharpness, so finding new sticks and keeping the edges sharp became part of their routine.

Many years or even a few generations followed; the use of sharpened sticks had become standard in all the groups now. Today, there was another group of hunters chasing a pack of reindeer. As they were all running, the leader of the group, Arjun, stumbled on a stone that was coming out of ground with sharp edges and immediately fell. When Arjun tried to see what happened, he saw that there was a deep cut on his foot. The trail of blood led to the piece of stone lodged in the ground with its sharp edge pointing up. He looked at it from all angles and then he started to dig around it. Finally, he was able to extract the whole stone out. It was definitely heavier than a stick but could do a lot more damage. Strangely enough, it had a hole on the other side. Arjun just put his stick into that and voila! It lodged quite nicely!! Now, Arjun had a much more potent tool than just a sharpened stick and also something that does not need daily sharpening. It was the start of the next generation of tools, and with that stone age!! The advantage with stones was that they did not lose their shape and were able to kill even faster and with higher precision. It was not easy to find similar stones for all the members of the group, but soon enough they were able to conjure up enough supplies.

The concept of stone tools spread to all the other groups, when they interacted with each other either in a friendly visit or on a battlefield. However, these tools still had the limitation that they were effective only in close range. When the target was outside of their direct reach, they were mostly useless. Then came another accidental discovery of elasticity of skin or sinew from dead animals or some form of thread from trees. Equipped with these materials, humans were able to make the next generation of tools such as a bow and arrow or a slingshot capable of attacking the prey from a much longer distance compared to direct reach. The accidental innovation and its spread continued.

Now, we can ask a question: why did such accidents happen only with humans? Are they just naturally lucky? Is some natural superpower playing favorites with them? Quite likely, not! All the other species also must have encountered such accidents, but they did not really learn from them, and even if that individual organism did learn something, it was not able to pass the knowledge to its fellow mates, leave alone the subsequent generations. However, only humans were able to learn from such accidents, remember those learnings, and pass on these learnings from one individual to another and even one generation to another, thereby improving on the learnings through time. What made this impossible for the other species? The answer to this question brings us to the final piece of puzzle with yet another skill that humans possess that is superior to all the other species, and that is our vastly superior communication abilities of humans compared to other species.

Humans show an uncanny ability to communicate their thoughts in an intricate manner with other humans in a way that other species cannot dream about. The highly developed vocal tract in humans is capable of producing a large array of complex sound patterns, thereby enabling humans to assign a large number of different meanings to each pattern, essentially enabling them to create languages. On top of oral communication, humans have also invented tools to capture their thoughts in a written manner, enabling them to preserve and pass the learnings from one generation over to the next generation. Combining the skills of communication with the ability to build and improve tools really catapulted humans into a whole new realm and now we can clearly see why humans are so much more advanced than all the other species. What dinosaurs achieved over tens and hundreds of millions of years, humans have achieved in matter of tens of thousands, occupying the entire Earth and dominating everywhere.

Conclusion

Equipped with this special ability of capturing and communicating information, humans have been building on top of their collective learnings (albeit accidental) and wisdom for millions of years, a feat that was not achieved by any other species. It is also fascinating to see that as humans' collective intelligence is growing, their rate of improvement is also increasing, leading to an exponential growth in progress. If we compare the amount of improvement we have achieved in the last century to what we achieved over the last hundred thousand years, the differences are staggering. It is absolutely mind boggling to imagine what lies ahead when we look at the future.

CHAPTER 5

Origin of Machines

"The machine has no soul, but it has a purpose," Arthur C Clark aptly noted in his book *Profiles of the Future* in 1962. Or a more dramatically worded version such as, "The machine is a reflection of our own desires and fears. It is a tool that we can use to create a better world, or to destroy it." By Isaac Asimov in his novel, *I, Robot* in 1950, try to describe the role of machines in our lives. Since their invention, machines have become an integral part of human life. Machines give humans superpowers; machines enable humans to do things that they can imagine but cannot make real. It is the single most important human invention that truly separated them from all the other species on Earth. No other species other than humans can create and use machines!

In the previous chapter, we looked into how the use of tools starts separating humans from other organisms. Machines mark the natural progression in the process of improving tools. Machine as such, however, is a fairly generic term and consequently can appear quite vague in its meaning. It can be used to describe a broad set of objects. Typically, when we refer to any object as machine, it involves some moving parts and is built to perform a specific function. A tool is just a handheld and manually operated device used to perform one or more tasks, while machines on the other hand consist of a complex assembly of parts that is powered by some energy source to perform a specific but a rather complex task. While a tool generally consists of a single part (or multiple parts connected in a rigid manner), a machine by definition consists of multiple components accompanied by moving parts. Examples of modern machines include engines, motors, electronic devices.

© Ameet Joshi 2024
A. Joshi, *Artificial Intelligence and Human Evolution*,
https://doi.org/10.1007/978-1-4842-9807-7_5

Examples of Early Machines
Wheel

While our current perception of machines revolves around such advanced concepts, machines need not be that complex by definition. **Wheel**[1] can be considered as one of the early machines built by humans. The earliest evidence of its use by humans dates back to about 3500 BCE. A very simple assembly of spokes and a circular body transformed the human race into a new era leaving all the other species to dust. Creation and use of machines marks the fundamental separation boundary between humans and all the other species. There exists plenty of evidence of other species making use of tools, but there is absolutely no evidence whatsoever of any other species making use of machines. Even when we consider the simplest of them all, a wheel, no other species other than humans have been observed to use it. One may observe some monkeys or crawling animals sliding over naturally occurring circular objects, but they do not understand their nature and cannot build one from scratch. You may see a trained chimpanzee riding a bicycle, but it did not learn to ride on its own and had been explicitly trained by humans, and building a bicycle is far beyond their capabilities.

Who was the first human to use a wheel can be up for debate, but it is generally accepted that around 3500 BCE, people in present day Egypt, Iraq (Mesopotamia), Russia (northern Caucasus region), and China were familiar with the concept and were routinely using it. Some of the earliest fossils point at the use of clay for building wheels. Such wheels were used in the manufacturing of pottery. Later, wood and stones were used in the creation of wheels and such wheels were used for building some form of

---

[1] A round disk made of single part can also be called a wheel, but that is not an example of machine, it is just a tool. But when we consider a wheel with multiple parts such as spokes and axle, it becomes a machine.

carts or trolleys for transportation. Humans then employed strong animals such as oxen and donkeys to pull these trolleys when loaded with heavy material. The use of such carts catapulted transportation of goods to the next level enabling exchange of goods across different regions. Over time, the design of the wheel evolved, with improvements such as the addition of spokes to make them lighter and more durable. The use of wheels and carts spread throughout the ancient world, and they were eventually used for a wide range of purposes, from agriculture to warfare to commerce.

Shaduf

Another interesting machine that was developed by ancient humans is called *Shaduf*, as shown in Figure 5-1. The origin of shaduf is not accurately known, but it is estimated that since 4000 BCE ancient civilizations in Egypt, Mesopotamia, and the Indian subcontinent were using it.

Figure 5-1. *Use of Shaduf by ancient humans to help draw water from river*

A shaduf, as depicted in the accompanying picture, involves a long pole or beam with a fulcrum or pivot point. We also call such an assembly a lever. The pivot point separates the stick into a longer segment and a shorter one. A heavy weight is tied to the end of the short segment, while an empty water bucket is tied at the end of the longer segment with rope.

This simple setup makes a nice use of mechanical advantage[2] emerging from the differences in the lengths of the arms. The rope tied to the bucket can then be pulled with hand to immerse the bucket in water. Once the water is filled, the weight on the other side automatically pulls the heavy bucket out of water and then it can be emptied into a farm to irrigate the crops or use the water for other purposes. Pulling such buckets of water without shaduf can be quite tiresome and may not be even possible if the bucket is too big or the level of farm is much higher than the level of water, but this simple assembly provided superhuman capabilities at the time and made the process of drawing water from river efficient by order of magnitudes. The use of the shaduf in ancient civilizations was a significant milestone in agricultural technology and played a crucial role in the development of early civilizations. The ability to irrigate large areas of crops allowed for increased food production that was needed with the population growth. This was a steppingstone toward the development of cities and civilizations. It was also used to move heavy objects, such as blocks of stone, and to lift building materials to higher elevations during construction.

Inclined Plane

Another curious example of a machine invented in ancient times is an inclined plane, as shown in Figure 5-2. An inclined plane may not appear as a machine as it does not have its own moving parts, but it facilitates directional change in the motion of the objects on top of it. As one pushes an object on an inclined plane in horizontal direction, the object is also pushed in vertical direction by the nature of inclined plane. As simple

---

[2] When two arms with different lengths are pivoted across a fulcrum, it can provide a system with mechanical advantage. The product of the size of the arm to the weight tied at the end of it is same for both arms. Hence, if a heavy weight is tied to the short end, it can be pulled with a weight much smaller by simply adjusting the ratio of the arms.

as it sounds, this machine had a huge impact on ancient civilizations. Construction of pyramids in Egypt around 2600 BCE would not have been possible without the use of inclined planes. Inclined plane reduces the force required to lift a heavy object by spreading the efforts over greater distance. The Egyptians used long ramps made of mud bricks to move the massive stone blocks from the quarry to the building site. The inclined plane allowed them to move the blocks uphill with less effort than if they had tried to lift them directly, which would have been impossible. The use of the inclined plane also became popular in other aspects of construction as well, where it was used to move heavy materials such as timber and Earth. In medieval Europe, the inclined plane was used to transport goods up and down hills. The rails of early railways were also used inclined planes, with carts or carriages being pulled up and down the tracks by horses and later by steam engines. Today, the inclined plane is still used in many applications, from wheelchair ramps to loading docks. The design of modern inclined planes has been refined to make them more efficient and easier to use, but the principle still remains the same.

Figure 5-2. *Ancient Egyptians using inclined plane to move large rocks to build pyramid*

Pulley

Pulley marks yet another jewel in the bucket of early inventions in machines. A pulley is a simple machine that consists of a grooved wheel and a rope or cable, as shown in Figure 5-3. It is believed to have been invented by ancient humans as a way to lift heavy objects with less effort. The exact origin of the pulley is unclear, as it was developed independently in multiple cultures over time. However, evidence suggests that pulleys were used as early as 4000 to 5000 years ago or around 2000 BCE to 3000 BCE in ancient Mesopotamia, Egypt, and Greece. The concept of the pulley likely originated from the observation of how ropes or cords could be used to lift objects by wrapping them around a fixed point, such as a tree trunk or a beam. By attaching a grooved wheel to that fixed point and running the rope through the groove, the wheel could be used to change the direction of the force applied to the load, making it easier to lift heavy objects. Early pulleys were likely made of wood, with the grooved wheel carved from a solid piece of wood or constructed from multiple pieces of wood held together with bindings or pegs. The rope or cord used with the pulley was typically made from natural materials such as plant fibers or animal sinews. Pulleys were used in various applications, such as construction, transportation, and agriculture. Pulleys were also used in the construction of monumental structures like the pyramids in ancient Egypt along with inclined planes to lift massive stone blocks. In Greece, pulleys were also used in theaters and temples to hoist heavy props or decorations onto the stage or roof. Pulleys were also used in ships to raise and lower sails, and in wells to draw water. Over time, the design and materials used in the construction of pulleys evolved. Metal pulleys were introduced, which were far more durable and efficient compared to wooden pulleys. Pulleys with multiple grooves or sheaves were developed, allowing for increased mechanical advantage, or the ability to lift heavier loads with less force. Different types of pulleys, such as fixed pulleys, movable pulleys, and compound pulleys, were also invented to suit different purposes.

During the industrial revolution in the 18th and 19th centuries, pulleys became a critical component in machinery and an integral part of the manufacturing processes. Today, pulleys are used in a wide range of applications, from elevators and cranes to conveyor belts and home exercise machines.

Figure 5-3. Use of pulleys by ancient civilizations to lift heavy objects

Screw

Another important machine that was invented in ancient times, albeit thousands of years after the invention of shaduf, and also was used for similar purposes of drawing water from natural reservoirs is called as Screw, or Archimedes' screw from Archimedes who invented it. Archimedes was a brilliant mathematician and engineer from Syracuse, Greece and created this contraption around 300 BCE. The device consists of a screw-shaped tube that is turned by a handle or crank as shown in Figure 5-4. When the tube is inserted into water, the screw lifts water from a lower level to a higher level. Archimedes' original design was likely used for irrigation purposes, allowing farmers to lift water from rivers or wells and transport it to their fields for watering crops. Over time, the screw was adapted for other uses, such as pumping water out of ships and mines and draining swamps. The use of the Archimedes' Screw spread throughout the ancient world and was later adopted by the Romans, who used the device extensively in their famous aqueducts and mines. The screw was also used by the Islamic world during the Middle Ages, where it was used for irrigation and water lifting. During the Renaissance, the Archimedes' Screw was rediscovered by European engineers and adapted for a variety of uses. The screw was used to power water mills and was later incorporated into steam engines during the Industrial Revolution. Even today, the Archimedes' Screw is still used in various applications, including wastewater treatment, hydroelectric power generation, and even some modern irrigation systems. The design of the screw has been modified and improved over time, but the basic principle remains the same. It's the ingenuity of its design and versatility that made it a timeless masterpiece in the history of machines.

Figure 5-4. *Showing the innards of Archimedes' screw helping irrigate by drawing water from river*

The use of machines was not restricted to driving mechanical benefits, but it was also used in science, specifically astronomy or even stargazing. In ancient times, due to the absence of city lights, the clear sky was always lit with stars and other celestial objects. The changing positions of these objects offered a great insight into a plethora of subjects.

Astrolabe

Stargazing held a profound significance for ancient humans, offering a multitude of practical, cultural, and spiritual benefits that enriched their lives in various ways. Before the advent of advanced navigation tools, the position and movement of celestial bodies, such as stars and planets, served as vital guides for navigation. Ancient mariners and travelers used the stars as celestial markers to navigate across land and sea, enabling them to chart courses and reach distant destinations with a degree of accuracy. The rhythmic passage of stars across the night sky provided early humans with a basic form of timekeeping. By observing the position of certain stars and their patterns of movement, they could track the passage of time, including the hours of the night when the Sun was absent. With longer term observations, it led to identifying the changing of seasons as well. Ancient agricultural societies relied heavily on celestial observations to determine seasonality and finding the most opportune times for planting and harvesting crops. The regular and predictable patterns of stars helped them anticipate seasonal changes, ensuring successful agricultural practices and food production. The celestial realm also had profound religious and cultural significance for many ancient civilizations. Stars and celestial events often played a central role in religious rituals, mythologies, and cosmologies. They were associated with deities, stories of creation, and the cycle of life and death, enriching the cultural tapestry of societies. Stargazing also influenced architectural and urban planning. Some ancient structures and cities were aligned with celestial events, such as solstices and equinoxes, reflecting a deep connection between the human-built environment and the celestial sphere. Observing celestial phenomena allowed ancient humans to predict certain natural events, such as eclipses and meteor showers. While these events might have been sources of wonder and sometimes fear, understanding and predicting them showcased a degree of scientific insight. Given such a deep-rooted role that the location of celestial objects in the sky was playing, machines were also invented to help improve the accuracy of these predictions. One such instrument is called the astrolabe, as shown in Figure 5-5.

Figure 5-5. *Astrolabe*

The astrolabe was used for solving problems related to time and the position of celestial objects. Its history again dates back to the ancient Greeks, who invented a primitive form of the instrument around 150 BC. However, the design and use of the astrolabe continued to evolve and improve over the centuries, as it spread to other cultures and regions of the world. During the Islamic Golden Age in the eighth century AD, Muslim astronomers developed a more advanced version of the astrolabe. They used this improved version to make highly accurate measurements of the positions of stars and planets to navigate the seas with better accuracy. By the medieval period, the astrolabe had become a common tool among European astronomers and navigators, who used it for everything from determining the time of day to charting the positions of the stars. Over time, the astrolabe continued to evolve, with new features and refinements being added to its design. By the Renaissance, it had become one of the most important instruments in astronomy and navigation and was used by some of the greatest minds of the time, including Copernicus and Galileo. Today, the astrolabe is no longer used for scientific purposes, but it remains an important part of the history of astronomy and navigation. Its elegant design and intricate workings continue to fascinate scholars and enthusiasts alike, and many replicas of the instrument can be found in museums and private collections around the world.

Timekeeping Machines

Along with complex machines like astrolabe, humans were also getting huge benefits from relatively simple machines such as sundials for timekeeping in day-to-day lives. Timekeeping is one of the most sought-after technologies that keeps us on track. We have made huge improvements in the accuracy of measuring the time, but it all started with just looking at how the shadow cast by the sun moves throughout the day. Sundials were then coupled with water clocks, also known as clepsydrae, where time was measured based on the amount of water that flows through predetermined pipes. These clocks could be used on rainy and cloudy days when Sun is not accessible as well as during nights. The accuracy of clepsydrae depended on temperature of water, the reliability of mechanical components and maintenance, etc., but overall marked a big improvement in timekeeping. The next milestone in timekeeping came with the invention of mechanical clocks around the thirteenth century AD. The mechanical clocks consisted of complex assembly of gears and weights and springs. With ever increasing sophistication, the accuracy of these mechanical clocks was exemplary at the time. Mechanical clocks got yet another boost in their timekeeping accuracy with the invention of the pendulum in seventeenth century AD. As the period of pendulum is constant to extremely high degree of accuracy (depends only on Earth's gravitation and length of the pendulum), the clocks based on this principle were extremely accurate. Depending on the material used for the pendulum, it could induce errors as the change in temperature changed the length of it. But the use of special alloys that do not change their dimensions with temperature or glass pendulum clocks represented the epitome of timekeeping till the twentieth century. Next biggest jump in accuracy came in the 1920s with the invention of quartz clocks. These clocks were based on the frequency of vibrations in a quartz crystal. As these vibrations are not affected by temperature as was the case with pendulums, they further improved the accuracy of timekeeping to the order of plus or minus 15 seconds per month. It still had some dependence on temperature and other factors such as the type of quartz crystal of the battery used to power the clock, etc. The next and the latest invention in timekeeping machines came through the use of atomic clocks that can boast an accuracy of plus or minus 1 second in over 100 million years! These clocks are based on natural resonance frequency of atoms (typically Cesium atoms). However, these clocks are huge and cannot be made into wristwatches. The wristwatches are still made with quartz technology or mechanical technology.

A careful look at the design and invention of all these machines points to a common thread across all of them: being able to perform things that we can think about, imagine, or contemplate and even "do" to some extent. Machines do not do anything that is not humanely plausible to a certain degree, but they enable humans to do those things easily, on a much higher scale, and with much less effort. Another important thing to note is that all the tasks that the machines do, they don't do by themselves. Humans are always involved in the operation. Humans are either using them, handling them, or powering them in some way.

The energy that is required by the machines to operate always comes from humans (directly or indirectly) and none of the work that these machines do is something brand new in principle. True, the amount of weight we could lift or the length of distance we could move the weight would be drastically reduced without the use of these machines; or the accuracy of predicting the location of stars would be rather poor without the use of machines; however, it is important to note that none of these machines really enabled humans to do something they did not know how to do without them. There was no magic, but they followed the basic principles of physics and provided a tunable mechanical advantage to humans. Also, any task that was achieved with these machines needed exactly the same amount of overall energy as it would have needed without their use; however, it is the way the energy could be channeled or modulated was changed with the use of machines and that alone was sufficient to achieve all the amazing engineering feats that humans were able to accomplish with them. All these ancient machines can be classified as purely mechanical machines. Not only did they make use of only mechanical energy, but they were entirely powered by humans or other animals.

Windmill

A deviation from this trend came in the form of a windmill, as shown in the Figure 5-6. A windmill was the first machine created by humans that made use of an energy source outside of human or any other biological entity. It harnessed the energy from the wind. It was quite an incredible feat. Wind energy was available for free essentially and using that energy to power a machine was nothing short of magic at the time. The credit of inventing the first windmill goes to Heron of Alexandria in Greece about 2000 years back. His ingenious design consisted of a chain of gears mounted in vertical direction and they would rotate as wind moved the top shaft. Wind power was ultimately harnessed to grind the grains, which was done with human power before. The design was not quite efficient and went through many revisions over the next several centuries to perfect it. The truly modern design of horizontal axis windmill was discovered in Persia (modern day Iran) around 1300 year ago or in ninth century AD. These windmills were used to grind grains as well as drawing water from wells.

Figure 5-6. *Ancient windmill employing power of wind to grind grains*

Waterwheel

Another interesting use of natural energy sources in machines can be seen in the creation of waterwheels. These wheels were laid out in the path of a river or stream that is doing down a slope. The flow of water would rotate the wheel generating energy. This energy was also used in similar applications such as grinding grains and fetching water. The origin of waterwheels is not quite as definitive as windmills, but oldest references of its use go back about 2000 years in ancient Rome and Greece. The technique was likely invented independently in many places.

Heat

Heat is another example of natural energy that can be obtained from the Sun or by burning wood or coal, etc. The first use of heat by humans dates back to prehistoric times, when early humans discovered fire and learned to harness its heat for cooking, warmth, and light. The exact date and location of the discovery of fire is unknown, but it is believed to have occurred around 1.7 million years ago. At that time, modern humans or *Homo sapiens* were not even born. These early humans belonged to the species we now call *Homo erectus*. These early humans used fire primarily for cooking food and for generating warmth during cold weather. As they became more skilled at controlling the fire, they also used it to produce light and used it for protection against other species as well as hunting. The discovery of fire and the harnessing of its heat were major technological advancements for early humans, allowing them to not only survive but thrive in harsh environments. However, humans were not able to benefit from the heat energy for anything other than its direct natural use for a very long time. In other words, humans could not muster the heat energy to power any type of machines. Fast forward more than 1.6 million years, and about 2000 years back, or in first century AD, in ancient Greece, a curious little gadget was invented called an **aeolipile**. The gadget consisted of a sphere and two pipes mounted on a circular structure, as shown in Figure 5-7.

Figure 5-7. *Aeolipile was the probably one of the earliest inventions that converted heat energy into motion*

The sphere was filled with water and heated over a fire, causing the water to turn into steam and escape through the pipes, which caused the sphere to spin rapidly. Although there is no evidence that this aeolipile was used for any practical applications, the machine certainly showed evidence of humans being able to use heat energy indirectly to move things through an intermediate step of heating the water to steam. However, it was not until eighteenth century AD in Britain when the concept was further refined, leading to the development of a steam engine where we were successfully able to convert the heat energy into something fundamentally different in the form of motion. In the early eighteenth century, the first successful use of steam engines was applied to pumping out water from coal mines to avoid flooding. The credit for this invention goes to Thomas Newcomen, whose initial concept of the steam engine was then further improved by James Watts. The improved design of James Watt's steam engine was much more efficient, and it powered textile mills, ironworks, and other factories. They were also used to power transportation, such as steam locomotives and steamships. Invention of steam engine was a key milestone in British industrial revolution.

Naturally occurring objects like wood were used to generate heat since the time of *Homo erectus*, millions of years ago, but it took a long time to identify fossil fuels like coal as even more potent sources of energy. The earliest documented evidence suggests that people in China had started mining and using coal to generate heat about 5000 years ago, while another evidence suggests that in ancient Greece, bituminous[3] coal was used as fuel for heating and cooking around 2300 years ago. The Greek philosopher Theophrastus described coal as a useful fuel in his work *On Stones*, written in the third century BCE. Chinese civilizations were also leading in the use of other fossil fuels such as natural oil and gas as energy sources about 4000 years ago or 2000 BCE. They even took a step forward to make clever use of naturally occurring bamboo stems as pipes to carry the gas directly into their homes. Once again, the use of all these fuels was still limited to generating heat and its direct use in cooking, generating warmth, etc. It was only during the eighteenth century in Britain that advancements in machine technology came together with the use of alternative sources of heat such as coal and other fossil fuels to drive the industrial revolution.

---

[3] Bituminous coal is black or dark brown in color and has a high carbon content. It is also relatively hard and dense, and it has a high heat content. Bituminous coal is naturally found in many parts of the world, including the United States, China, Russia, and Australia.

Natural oil is an even more potent source of naturally occurring energy source with about twice the energy concentration compared to coal. Humans were aware of natural oil for many centuries. Earliest use of natural oils goes back to 600 BCE in China, where it was used to fuel lamps. A larger scale use is observed in the ninth century along the border of Europe and Asia or modern-day Azerbaijan, where first oil wells were created, and natural oil was extracted as fuel. However, first commercial use of natural oils started as late as early nineteenth century when Americans started drilling wells and the oil and gas industry was born. The use of different types of energy sources has played an important role in the advancement of machine technology. With the advancements in technology, the focus was on using more and more potent energy sources.

Electricity

The use of mechanical or heat energies certainly put us on a fast track toward advancement, but electricity marked an entirely new era in the use of energy to drive machines. It was the first time the energy was completely invisible to the naked eye. In earlier times, we could see mechanical gadgets in action, we could see flames and things getting heated, but electricity was a whole new beast. Electricity was not entirely unknown to humans till the modern times, but it was the least understood as a source of energy. There is evidence that can be traced back to ancient times around 600 BCE, when Greek philosopher Thales of Miletus had observed that rubbing amber (a fossilized tree resin) could attract lightweight objects like feathers. Although this was a direct application of what we now know as static electricity, at the time, it was nothing short of magic. Also, we could not really make use of this observation as a source of energy in any meaningful manner. It took over two thousand years until the seventeenth century that we slowly started to grasp the nature of electricity. Around 1600 AD, experiments involving static electricity were conducted by scientists such as William Gilbert and Otto von Guericke.

They explored the properties of electrical attraction and repulsion. Gilbert is also credited for coining the term "electricity" to describe this force. In the eighteenth century, Benjamin Franklin, a brilliant scientist, conducted a rather daring experiment that advanced our understanding of electricity. He was the first human to harness electricity directly from a thunderstorm. It is a fascinating as well as awe-inspiring story.

Story of Benjamin Franklin

Amidst the stormy skies of 1752, Benjamin Franklin embarked on an audacious quest to unravel the secrets of the heavens. Armed with his intellect, a sturdy kite made from silk, he embarked on an adventure to alter our understanding of nature's most unique force. The air crackled with anticipation as Franklin meticulously assembled his apparatus. His kite was decked with a silk handkerchief, and at the apex of the kite, a slender metal wire pointed skyward, poised to capture the very essence of the turbulent heavens. Clutching the silk string with one hand and grasping a silk ribbon to insulate himself with the other, he emerged from his shelter, stepping into the chaotic embrace of the storm. Franklin, undeterred by the commotion around him, unleashed his kite into the raging heavens. It soared into the darkness, a tiny vessel sailing amidst the wrath of nature, bridging the gap between the mortal realm and the unknown. As the lightning bolts streaked across the sky, Franklin's eyes gleamed with both excitement and trepidation. He was somewhat aware of the risks involved, yet his thirst for knowledge propelled him forward. With bated breath, he waited, feeling the very weight of the universe upon his shoulders. Then, in a flash of brilliance, it happened. A bolt of lightning, mighty and fearsome, cleaved through the darkened firmament, hurtling toward the Earth with a resounding crack. Franklin's kite became a conduit for the celestial power that coursed through the heavens. The world seemed to hold its breath as Franklin observed the remarkable effects of his daring experiment. And then, as if destiny itself beckoned,

Franklin cautiously reached out his hand, extending a finger toward the dangling key. With a surge of electrical energy, a spark leapt from the key to Franklin's waiting finger. The shock reverberated through his body, an electrifying jolt that marked the convergence of science and bravery, forever etching his name into the annals of history. In that thunderstorm, amid the torrents of rain, Benjamin Franklin unveiled the profound truth: that lightning, the harbinger of destruction and wonder, was an awe-inspiring manifestation of the captivating power we now know as electricity.

The details of the experiment can be debated, but the result of the experiment is undisputable and established electricity as a definitive source of energy. The next crucial step in harnessing electrical energy came through the invention of Alessandro Volta who invented the first electric battery in 1800. This development provided a steady source of electricity that can be used whenever and wherever it was needed. In the nineteenth century, scientists made great strides in understanding the nature of electricity. Thomas Edison invented the incandescent light bulb, which made it possible to use electricity for lighting. Nikola Tesla developed alternating current (AC) electricity, which is the type of electricity that is used in homes and businesses today. All these inventions changed the way we power machines quite dramatically, thereby changing the world.

Electricity, as it was discovered and used in the early days, was strictly through the wires. However, humans were also experimenting with some of the effects of electricity that could be observed without the use of wires. Specifically, in 1666 AD, an English physicist Robert Boyle first observed that a spark of static electricity can produce a disturbance in a beam of light that is felt a few feet away. He was not able to explain how or why this was happening, but he did record his observations. Another key discovery in the field came around early eighteenth century when Danish physicist Hans Christian Orsted discovered that electricity and magnetism are somehow related. In his experiment, he observed that if you take a wire that is carrying current, it can move a magnetic needle nearby. This discovery paved the path for the emergence of the field of electromagnetism. Around the same time, a French physicist Andre-Marie Ampere was uncovering more details of this interaction. He was the first who established the quantitative relationship between the electric and magnetic fields by creating his famous law, aptly called Ampere's law. Later in the century, another famous English physicist Michael Faraday proved the other type of interaction between electricity and magnetism where a changing magnetic field can produce an electric current in a wire through the phenomenon of induction. His work laid out a solid foundation for the practical applications of electromagnetic principles. Generation of electricity using dams or wind turbines was possible only through the principle of induction. Induction also led to the discovery of electric motors establishing electricity as the most accessible as well as the cleanest source of energy in our lives. Generation of electricity did not create any unpleasant side effects such as smoke. Scottish mathematician and physicist James Clerk Maxwell truly integrated all the pieces of experimental information regarding the interactions between electricity and magnetism and published his renowned theoretical work on electromagnetism, thereby establishing it as a formal scientific area of study. Around the 1860s, he formulated a set of equations deeply rooted in mathematics of calculus that established the ultimate reference for quantitatively understanding and predicting the effects of all electromagnetic interactions. Most of the electromagnetic interactions that scientists were studying at this time were wireless but restricted to close vicinity of objects, typically ranging from a few centimeters to a few inches. However, Maxwell's equations also predicted the existence of electromagnetic waves that could travel in theory much longer distances. In the late 1880s, another German physicist Heinrich Hertz conducted a series of experiments where he could demonstrate a type of electromagnetic waves we now call as radio waves that could travel a distance of a few meters. His experiments laid the foundation for long distance wireless communication, building on which subsequent researchers, such as Guglielmo Marconi, were able to extend the range of radio wave transmission over significantly longer distances, making radio communication possible.

Invention of electricity as an extremely sophisticated source of energy coupled with the use of electromagnetic waves for wireless transmission catapulted humans into the next era of machine technology. The effect of these inventions was so vast and groundbreaking that the differences in human life between the eighteenth and nineteenth centuries were larger than differences between human life from eighteenth century extending all the way back to ancient times in BCE. The power of electromagnetism and its pinpoint control was unlike any other source we had used before.

Printing Machine

Improvements in the use of machines and the energy sources that powered them went hand in hand for centuries. Another type of machine that sits at the very core of human advancement is the ability to write on a large scale, or in other words, printing technology. The evolution of printing technology has been a remarkable journey, transforming the way humans communicate, share information, and disseminate knowledge. The ability to effectively share information through the means of printing has been the key factor driving human technological advancement. From ancient methods like woodblock printing to modern day digital printing, this evolution has shaped societies, cultures, and the global exchange of ideas.

Woodblock printing, with its origins in ancient China, involved carving text or images onto wooden blocks. Ink was applied to the carved surface, and the block was pressed onto paper or other materials. This method was used for centuries to produce books, manuscripts, and artworks. It was rather a rigid technique and could not adapt to changing content quickly. This improvement came by way of movable clay type printing. It was invented again in China in the eleventh century, where individual characters could be arranged and rearranged for printing. Johannes Gutenberg's introduction of metal movable type in the mid-fifteenth century revolutionized this technique. Gutenberg's mechanical printing press marked a turning point in human history. The new printing press could produce books more efficiently than manual methods. This innovation played a crucial role in standardizing language, disseminating ideas, and accelerating the spread of information. The Industrial Revolution brought mechanization to printing. Steam-powered printing presses enabled large-scale production of newspapers, books, and other printed materials. This period saw the rise of publishing industries and mass media. The invention of offset printing allowed ink to be transferred from a plate to a rubber roller and then onto paper. This technique improved print quality and enabled faster production, becoming a standard method for many types of printing. Photolithography introduced photography into the printing process in the twentieth century. Images could be transferred onto metal plates using light-sensitive chemicals, enhancing print quality and visual representation. The notion that a picture is worth a thousand words came to reality with this technique. In the late twentieth century, the digital era revolutionized printing. Digital printers use computer files to produce copies directly, eliminating the need for traditional printing plates. This method allows for customization, on-demand printing, and reduced waste. This technology brought the printers into our households where, in earlier times, one needed a huge building to accommodate a printing press. The relentless advancements with computers brought another game-changing milestone in printing technology in the form of 3D printing. It allows the creation of physical objects layer by layer from digital models. Now, we can not only replicate the text and images, but we can actually create entire objects based on their model. Along with advances in the materials used in printing beyond traditional ink, such as plastics, metals, ceramics, and even biological materials, boundaries between printing and engineering are all but vanishing.

Similar to printing technology that enabled the mass production of written materials, textile industry revolution enabled mass production of garments, thereby tackling the challenge of clothes manufacturing for the growing human population. Creating protective clothing has been one of the fundamental requirements of humans since ancient times. It started with the use of naturally occurring materials such as leaves and hides from dead animals. These materials were used to make handcrafted clothing artifacts for thousands of years. The use of fibers along with machines such as spindle and looms marked the beginning of the textile industry. The first use of these machines dates back to ancient times, but then these machines were strictly handheld devices with rather limited scope in quantity that could be manufactured. Textile industry stayed stagnated to these methods all the way till the industrial revolution and the birth of the steam engine. However, since then, it has taken huge strides in the form of inventions such as Spinning Jenny, Water Frame, and Power Loom that catapulted the garment production to production of complex patterns and accelerated the production rate. The twentieth century saw yet another milestone in the textile industry in the form of sewing machines and its various advancements ultimately leading to computer-aided textile production. These developments in machines and manufacturing technology also created the new concept of production lines, or assembly lines, where the manufacturing process involves a complex sequence of operations to be performed one after another in quick succession to enable efficient mass production of goods. This process breaks down dauntingly complex operations into a set of specialized micro-operations that can be performed by different machines aided by different individuals. This concept was later on further optimized for the production of automobiles and airplanes and so on.

The culmination of the wheel for the purpose of transportation was realized in the twentieth century with mass production of automobiles. Some of the early contraptions of automobiles used steam engines at the heart to power the motion, but soon it was replaced with internal combustion engines, as shown in Figure 5-8, with the use of natural oils. Internal combustion engines were truly the hallmark of the innovation in technology when they were first invented. The complexity and efficiency of its operation was off the charts. The first internal combustion engine was invented in 1860 by an engineer named Jean Joseph Etienne Lenoir. It was a four-stroke engine and used a mixture of hydrogen and oxygen to power it. In the years that followed, there were many improvements to the internal combustion engine. In 1876, Nikolaus Otto invented the four-stroke engine that is still used in most vehicles today. In 1885, Karl Benz invented the first gasoline-powered car. And in 1897, Rudolf Diesel invented the diesel engine. Natural oils were far more energy efficient, and their combustion could be better controlled for the purpose of adjusting the speed of motion. Henry Ford is typically associated with the first mass production of gasoline-powered automobiles and introduction of assembly lines along the way, although the first internal combustion engine powered automobile was created by Karl Benz in late nineteenth century in the form of Motorwagen. The evolution of cars from that point onwards is quite remarkable. The twentieth century brought more innovations into the design in the form of better aerodynamics, better fuel efficiency, better suspensions, brakes, transmissions and the list went on and on. Later on, electronics and computers were also integrated into the mix to enable onboard music, navigation, and driver assist features and so on. With increased speed came increased risks of accidents and that pushed the technology to build better safety features such as adaptive seat belts, airbags (which are essentially balloons that are deployed in strategic locations to protect the individuals inside the car when the car senses an impact indicating an accident), anti-lock braking systems, stability controls, and so on. Cars became a single instrument that showcased an entire gamut of human advancement in machine technology. In spite of being on the cutting edge of technology, it took a relatively long time for the cars to get their power from electricity. The reason was lack of sufficient storage of electricity that is compact enough to be hosted on the cars. However, in the twenty-first century, battery technology advanced enough to be able to support a reasonable range for cars to operate entirely on its power. This is marking as one of the biggest milestones in the history of cars, where all the manufacturers worldwide are migrating to electricity-powered cars from the longstanding gasoline or petrol and diesel-powered internal combustion engines.

Figure 5-8. *Internal combustion engine*

Nuclear Energy

Another important discovery in search of new sources of energy came in the early twentieth century in the form of nuclear energy. Nuclear energy provides the most powerful and concentrated source of energy ever known to humans. At its heart, nuclear energy is generated when a tiny amount of mass gets converted into pure energy. Although Einstein's famous equation $E=mc^2$ established a theoretical relation between mass and energy around the start of the twentieth century, he could not demonstrate any direct mechanism to actually carry out this conversion. To understand the details of this conversion, we need to take a step back into the physics of atoms.

The pioneering research by scientists such as Ernest Rutherford, Marie Curie, and others revolutionized our understanding of the atom. Rutherford's experiments led to the discovery of the atomic nucleus, a tiny, dense region at the center of the atom containing positively charged protons and uncharged neutrons. This model of the atom laid the foundation for further investigations into atomic processes. Another critical development was the discovery of radioactivity. In the late nineteenth and early twentieth centuries, scientists like Henri Becquerel, Marie Curie, and Pierre Curie studied the emission of radiation by certain elements, such as uranium and radium. They found that these radioactive materials spontaneously emitted subatomic particles and energy. The idea of releasing energy from atomic processes gained further significance as researchers began to unravel the mysteries of the atomic nucleus. Einstein's equation was guiding the amount of energy that could be released in this process. Another breakthrough came in the 1930s with the discovery of nuclear fission. In 1938, Otto Hahn, Fritz Strassmann, Lise Meitner, and Otto Frisch conducted experiments in which they bombarded uranium atoms with neutrons and observed the splitting of the atomic nucleus. This process, known as nuclear fission, released an enormous amount of energy, significantly more than naturally occurring radioactive emissions. With the discovery of nuclear fission, we truly began harnessing the power of mass. Further research by scientists such as Enrico Fermi and Leo Szilard led to the development of the first controlled nuclear chain reaction in December 1942 as part of the Manhattan Project. This chain reaction, which occurred in a controlled environment, demonstrated the potential of nuclear energy as a new, sustainable, and legitimate source of energy. The heat generated in fission could be used to generate electricity. It was vastly more powerful than other established sources such as dams that generated electricity, but it did come with its own issues in the form of handling of nuclear waste. Generation of electricity using water and wind were still the cleanest form of energy with no harmful by-products.

Computer and Robots

With electricity established as the most accessible source of energy and dams and wind turbines and nuclear powerplants as its producers, the twentieth century saw a rapid rise in machines powered by electricity and electromagnetism. First incandescent bulbs replaced the age-old candles to light up all the homes; electric fans followed by electric heating and air conditioning machines helped with managing the temperatures in our homes. However, the true epitome of machine technology came in the latter half of the century with the invention of computers. A new age of industrial revolution called the information age began. Computers started off as giant machines occupying entire buildings with rather mediocre computing capabilities in the 1950s, but soon, with the miniaturization of silicon chips and advancements in microprocessor technology, the computers became smaller and smaller in size and started offering more and more compute performance. Soon, all the paper-based information gathering, and storage systems were replaced with computers. Computers offered extremely accurate and reliable infrastructure with ease of access. Functions such as searching for information, sorting information based on one or more criteria, classifying and categorizing information that took long manual efforts became almost automatic. All that was needed was to type a few keywords and a few clicks of a mouse button and results were presented on the screen in a matter of seconds. The initial impact of computers was primarily observed in the areas that needed writing and reading, but the next invention that was heavily based on the computer technology changed that. The invention of robotic machines has been a significant technological achievement with far-reaching impacts on various industries and aspects of society. The history of robotic machines can be traced back to the mid-twentieth century, when the concept of a machine that could perform tasks autonomously or semi-autonomously was first realized.

The term "robot" was coined by Czech playwright Karel Čapek in his 1920 play *R.U.R.* (Rossum's Universal Robots), describing artificial, human-like beings created through manufacturing. However, the modern concept of robotic machines began to take shape in the 1950s and 1960s. The first industrial robot, known as the Unimate, was developed by George Devol and Joseph Engelberger in the late 1950s. It was installed in a General Motors plant in 1961 to perform tasks like welding and handling heavy materials. This marked the beginning of the use of robotic machines in manufacturing. It is also interesting to note that this robot did not look anything like a human. It contained a big assembly of a single arm and a large box underneath carrying all the necessary mechanical and electronic components.

Impact of Robotic Machines

Robotic machines changed the way manufacturing worked in earlier times. Use of robots in production lines increased their efficiency, precision, and speed in tasks such as assembly, welding, painting, and packaging by order of magnitude. This has led to reduced production costs, improved product quality, and the ability to manufacture complex products. In some industries, robots have taken over repetitive, dangerous, or physically demanding tasks, allowing human workers to focus on more complex and creative aspects of their jobs. Robotic machines have also found applications in medicine and healthcare, such as in surgical procedures. Robotic surgical systems enable surgeons to perform minimally invasive procedures with enhanced precision and control, leading to reduced patient trauma and faster recovery times. Robotic machines have enabled exploration in environments that are challenging or hazardous for humans. Space rovers like the Mars rovers have provided valuable insights into the Martian surface, while underwater robots explore deep-sea environments and gather data for scientific research. Robotics has also changed the way we used to do farming. Agricultural robots, often referred to as agribots or agri-robots, are being developed to perform all the major farming tasks such as planting, harvesting, and monitoring crops. These machines have the potential to increase agricultural efficiency, reduce waste, and address labor shortages. Robotic machines can also assist individuals with disabilities or limitations, enhancing their quality of life.

With advancement in silicon manufacturing, following Moore's law,[4] all the computer and even robotic infrastructure was going through miniaturization in the latter half of twentieth century and continued with even more vigor in early twenty-first century. The size of computers reduced from giant machines occupying entire floors to rectangular boxes smaller than a couple of feet in dimension as Desktops to the dimension of an oversized book in the form of Laptops ultimately down to handheld devices measuring just a few inches. All the while when the size was reduced, the computation power increased, making the smaller devices even more powerful and also requiring less and less energy to operate. The invention of smartphone was the true culmination of computer technology around the last decade of the twentieth century. A device that can fit in anyone's hand and pocket can record high quality photos and videos, capable of making wireless calls across the world, capable of sending and receiving emails with multimedia, capable of playing video games with high resolution 3D graphics, capable of performing complex neural processing, and the list goes on and on. Smartphones in the twenty-first century replaced automobiles from previous century as the de facto machines that showcase all the latest technology. They ushered in the new era of computation and social networking with the speed of light.

---

[4] Moore's Law is a well-known observation and prediction made by Gordon Moore, co-founder of Intel Corporation, in 1965. It refers to the exponential growth in the number of transistors that can be placed on an integrated circuit (IC) or microchip over time. This increase in transistor count has been accompanied by a decrease in the size of transistors and a corresponding improvement in computing power.

Conclusion

Human intelligence created the tools and later machines that separated us from all the rest of the organisms. In the last couple of chapters, we saw how these innovations progressed and led us to the path of ever-increasing sophistication in technology. All the innovations made our lives easier, happier, and safer. They did come with their own drawbacks and humans have been learning how to avoid them. The machines augmented human natural abilities to make them superhuman, but still, they could not function on their own. In the next chapter, we are going to look at how humans were able to instigate proactive intelligence into the machines and take the technology to the next level.

CHAPTER 6

Machine Intelligence

Machines started their journey with humans as an upgraded version of rudimentary tools such as sticks as we saw in the previous chapter. Very quickly they established themselves as an invaluable partner of humans in their daily life. They reduced the hardships from human life, they reduced the amount of hazardous work from human life, they enabled humans to carry out more than what they could do with their body alone, thereby making their lives safer, easier, more rewarding, peaceful, and ultimately happier. However, these benefits were not without challenges and accompanied dangers. These machines could also be used by humans against each other, they helped create ways of destroying human life at an unimaginable scale as illustrated by their use in world wars. The machines that save labor can also create more work; the machines that grant us more power can also rob us of control; machines that can help us live longer can also be the tools of our own destruction. Machines essentially create an extreme version of humans in nearly every aspect possible. In the earlier chapter, we saw the emergence of machines; in this chapter, we are going to see how they acquired intelligence.

Technological progress that spanned over the past thousands of years enabled the machines to gain more and more power and skills; however, they remained as dumb as a rock when left on their own or used outside of the rather narrow scope of their operation. The intelligence of machines arose as an entirely new concept only in the late twentieth century. Machine intelligence has always been a subject of incredible interest and

© Ameet Joshi 2024
A. Joshi, *Artificial Intelligence and Human Evolution*,
https://doi.org/10.1007/978-1-4842-9807-7_6

fascination to us since then. To really understand the origin of intelligence in machines, we will digress a little and look at what lies at the core of the operation of machines.

Classic Machines

Let's take the example of a windmill.[1] When the wind moves the fans of the windmill, the shaft connected to them rotates and it drives the metal flywheels, thereby grinding the grains. The faster the wind, the faster the shaft rotates, rotating the flywheels faster and effectively grinding the grains faster. If the wind is slower, grinding will be slower. Also, when there is no wind whatsoever, the windmill is going to stand still, and no grains are going to be ground. This operation is fairly straightforward. However, what if by mistake a stone gets into the grains and gets stuck in the flywheels? In that case, the whole windmill would stop operating. There is a very simple solution to this problem in the form of removing the stone, which likely requires much less effort than what is required for the whole grinding operation. However, the machine cannot figure it out by itself. If, on the other hand, there were a human (let's call him Adam) doing the same operation, and the flywheels got stuck with a stone, Adam would quickly notice it, remove it, and resume the grinding. Sure, Adam would be doing the job at a much slower rate, but he will not get stuck even if there are unforeseen interruptions. However, the machine, with the size and complexity of a windmill, would just go belly up and stop functioning with a minor interruption that is not anticipated in its design. What is it that Adam is doing differently in this case? Let's break down Adam's behavior. There are three distinct steps: (1) In the first step, Adam would observe the rock that is stuck in the flywheels using his eyes. (2) In the second step, Adam would stop using his hands to grind and instead

---

[1] Feel free to look at the figure provided in the earlier chapter for reference.

use them to pick up the stone that is stuck in the flywheels. These are the same hands that are rotating the flywheel, but the operation of removing the stone is quite different and requires a different skillset. (3) In the third step, Adam resumes the regular operation of the mill by switching the skillset of his hands. This three-step procedure involves what we call a feedback mechanism. Humans and all the living species for that matter are using this concept of feedback all their lives since birth by interacting with the environment and are almost oblivious of the complexities of it. This feedback cuts across all the sensory and reactive organs of humans and it comes natural to humans to acquire feedback through eyes or ears or touch and react to it with hands or feet or mouth and so on. However, machines are quite behind humans in this aspect. First of all, machines like windmills do not have any sensory parts on them that can sense the environment. They are typically composed of only reactive parts. Building a machine that can somehow sense the surroundings and then react to it turned out to be far more difficult than building ever more complex machines over the past thousands of years.

Mechanical Feedback

The first breakthrough achievement in this field of building machines that can somehow react to its environment and adjust its process came around the late eighteenth century when James Watt built his version of steam engine during the industrial revolution in Britain. It was not just a major improvement in steam engine technology, but it was a monumental leap for humankind that took us into a new era of smart machines that had the concept of feedback and self-regulation mechanism built into them.

In its basic form, as shown in Figure 6-1, a steam engine converts heat energy from burning coal into mechanical energy powering various machines and ultimately driving the vehicle.

Figure 6-1. *James Watt's steam engine with mechanical feedback system*

It consists of a boiler as shown in the picture at the left bottom that heats the water to produce steam, which then expands in a cylinder, pushing a piston. The reciprocating motion of the piston is transferred to a crankshaft located at the top, converting it into rotational motion. This rotational power is harnessed for applications such as factories, transportation, and mining. However, with variation in the heating process causing variations in the generation of steam, the output of the engine can become unpredictable. To overcome this, Watt introduced a feedback mechanism known as centrifugal governor as shown at the middle of the top just below the crankshaft, which enabled automatic regulation of the speed of the engine. The governor consisted of a rotating spindle, which was connected to the engine's drive shaft, and two weighted arms were attached to the spindle. As the governor operated, the spinning motion of the shaft caused the governor spindle to rotate. The weighted arms of the governor were positioned perpendicular to the spindle and would move outward or inward based on the rotational speed of the engine. When the speed increased, the centrifugal force caused the arms to move outward, and when the speed decreased, they moved inward. The movement of the governor arms was connected to a mechanism that controlled the engine's throttle valve. The throttle valve regulated the flow of steam to the engine, thereby controlling the power output. The connection between the governor arms and the throttle valve was typically achieved through a linkage system involving levers and rods. Here are the four steps that enable the feedback mechanism:

> *Step 1*: As the engine's speed increased beyond the desired level, the centrifugal force caused the governor's arms to move outward. This motion was transmitted through the linkage system to the throttle valve.

> *Step 2*: The outward movement of the governor arms caused the throttle valve to close partially or completely. By reducing the flow of steam, the engine's power output was reduced, thereby slowing down the engine.

> *Step 3*: Conversely, if the engine's speed decreased below the desired level, the governor arms moved inward due to reduced centrifugal force. This inward movement was again transmitted to the throttle valve.

> *Step 4*: The inward movement of the governor arms caused the throttle valve to open, allowing a greater flow of steam into the engine. This increased the power output and helped restore the speed of the engine.

Following a systematic approach like this, the centrifugal governor provided an automatic feedback loop that continuously adjusted the throttle valve based on the engine's speed. If the speed deviated from the desired setpoint, the governor would respond by modifying the steam flow, thus maintaining a relatively constant speed. The centrifugal governor was a critical innovation in Watt's steam engine as it offered a self-regulating mechanism for controlling the engine's speed. This feedback mechanism ensured that the engine could adapt to variations in load and maintain a stable and efficient operation, contributing to the overall success and widespread adoption of Watt's steam engine during the Industrial Revolution.

The sensors used in this feedback mechanism were not nearly as advanced as the ones that humans have at their disposal; however, these primitive and strictly mechanical sensors in the form of centrifugal governor arms were some of the first baby steps toward building a machine that can respond to changes in the surroundings and adjust its process.

Strictly speaking, Watt's steam engine was not the first machine to use some form of feedback. There are some examples in history going back thousands of years, where such mechanisms were used. In ancient Chinese Han Dynasty, about 200 BCE, a chariot was built with a figurine using an elaborate mechanical gear assembly, as shown in Figure 6-2. Some form of mechanical feedback system was used to ensure that the figurine on the chariot always pointed South regardless of how the chariot takes turns.

Figure 6-2. *Chinese chariot with a figurine that always points south*

Another notable example is the ancient Greek Antikythera Mechanism. This device was constructed around 100 BCE and used a complex mechanical assembly to perform astronomical calculations and predictions. It was used to model the movements of celestial bodies and track astronomical phenomena. However, these were fairly solitary examples and the technology developed for these devices did not progress in creating machines that actually made a significant difference in human lives. Leonardo da Vinci is considered as one of the greatest scientists, engineers, architects, and artists to ever see the face of Earth and unsurprisingly, he also played a role in creating feedback enabled machines. He had invented an improved version of a waterwheel that could maintain its rotational speed in spite of variations in the speed of flowing water in the stream using a feedback mechanism. The wheel featured a series of hinged paddles or buckets that would fill the water as they dipped into the flow. As the buckets filled with water, their increasing weight caused them to tilt downwards. This movement, in turn, activated a system of levers and gears that adjusted the position of the wheel to compensate for the increased load. By doing so, the mechanism regulated the waterwheel's speed and maintained a relatively constant rotational speed despite variations in the water flow. This mechanism in principle was quite close to Watt's steam engine; however, there is no evidence that Watt copied da Vinci's design. Da Vinci's design for the self-regulating waterwheel was part of his extensive collection of sketches and ideas, showcasing his inventive and creative prowess. However, there is no evidence to suggest that the specific design for the self-regulating waterwheel was ever constructed or employed on a practical scale. It is worth noting that da Vinci's contributions to engineering and his forward-thinking designs had a significant influence on later generations of inventors and engineers. While not all his designs were implemented during his time, they served as inspiration for subsequent advancements and innovations in various fields. From the perspective of present-day engineering, Watt's implementation still marks the true beginning of feedback-powered machines. Later on, this technology saw its culmination in the form of internal combustion engines that are used in most gasoline or diesel-powered cars in the twentieth as well as twenty-first century.

Even though machines such as Watt's steam engine did utilize feedback to improve or self-adjust their own functioning, the process was predefined at the time of conception of the machine. All the operational modes of the machines and all the possible variations of responses that could be experienced by the machine were already taken into consideration and the feedback mechanism was designed to address those variations and respond to them as desired. It was the genius of the creator of the machine that really mattered and once the design was implemented, it would keep functioning as expected for the life of the machine. However, if the machine faced a change that was not anticipated by the creator, the machine would still fail. For example, in case of Watt's steam engine, if the gears got clogged with dirt, the governor mechanism would not function as desired and the engine would fail; or if the engine is loaded with too much stress that is not anticipated, the engine would still fail.

To draw a critical comparison between the predefined efficiency of machines and the remarkable adaptability of humans, let's delve into a real-life scenario that unfolds with intriguing depth. Imagine a time when a team of genuine individuals was enlisted to undertake a task that is commonly delegated to an engine, yielding a narrative that resonates with the essence of human endeavor. In a quaint village nestled at the foothills of a majestic mountain range, there lived a close-knit community renowned for their unity and resourcefulness. The verdant landscapes and winding paths served as the backdrop for an intriguing experiment involving the village's own inhabitants. Meet Jacob, a robust and amiable man known for his strength and camaraderie. He was chosen to lead a team of villagers in pulling a cart laden with essential supplies to a neighboring village. Normally, such tasks were entrusted to a mechanized engine, which boasted unparalleled brute force and speed. However, this time, the village council had a unique proposition – to explore the dynamics of human adaptability. Accompanying Jacob were Emily, a young and spirited woman known for her innovative thinking, and Benjamin, an elderly sage renowned for his wisdom and resilience. The three, each representing a different facet of human potential, embarked on their journey with a mix of excitement and uncertainty. As they set out, the trio was armed with a basic understanding of the distance they needed to cover and a rough idea of the terrain they would traverse. Their journey commenced with the sun illuminating the path ahead, casting a sense of optimism. However, the vagaries of nature were poised to test their mettle in ways they could scarcely anticipate. Several miles into their journey, a sudden shift in weather caught them off guard. Dark clouds gathered ominously and rain began to pour in a torrential downpour, something that the village had never seen before. The once-firm roads morphed into treacherous quagmires, bogging down their progress significantly. In this unforeseen predicament, the raw power of the human team paled in comparison to the engine's might, but the might of human innovation was unparalleled. Amid the deluge, Emily's inventive mind sprang into action. Recognizing the need to adapt swiftly, she scanned their surroundings for shelter. Her eyes fell upon a colossal oak tree, its branches outstretched protectively. With Benjamin's sagacious endorsement, they huddled beneath the sturdy canopy, seeking refuge from the elements. Time passed, and the rain eventually relented, paving the way for the trio to continue their journey. However, the paths they had initially charted were now obscured by mud and muck. Here, the spirit of collaboration was brought to the forefront. Jacob's sinewy arms worked in tandem with Emily's strategic insight, and Benjamin's indomitable resolve, as they navigated through the altered landscape. With an unwavering spirit, the team pressed onward. Their conversations were filled with laughter, stories, and shared wisdom, forging a bond that transcended the challenges they encountered. Despite the setback caused by the capricious weather, their collective resilience shone through. Though they arrived at their destination slightly later than anticipated, the significance of their achievement was undeniable. The villagers who had observed their journey marveled not only at the supplies they brought but also at the indomitable human spirit they embodied. The journey of Jacob, Emily, and Benjamin became a tale passed down through generations, a testament to the multifaceted essence of humanity.

In the presence of routine and anticipated weather, a steam engine powered train would have completed this journey far more efficiently than the human trio, but when things went south beyond the realms of predictability, machines were outclassed. This story serves as a reminder that while machines possess remarkable efficiency and power, they lack the unique ability of humans to adapt, collaborate, and innovate in the face of the unknown.

So, what is it that makes the group of humans superior to a well-designed machine? It is through the process of continuous learning through cross-modal interactions! Before diving into the concept of cross-modal or multimodal interactions, let's focus on the aspects of learning. Learning is a fundamental process through which humans acquire knowledge, skills, behaviors, or attitudes, resulting in a permanent change in their behavior or mental processes. It is a complex and multifaceted phenomenon that occurs throughout our lives and allows us to adapt to our environment, solve problems, and improve our understanding of the world. Learning can take place through various mechanisms and in different contexts.

Concept of Learning

Although the scope of learning is rather complex, the fundamental premises on which it is based are quite simple. Let's define the learning process with a direct and simple approach. We first need to establish a few artifacts. The first artifact is an entity that is trying to learn, the second artifact is everything that is surrounding that entity; let's call it an environment. The entity that is trying to learn must be capable of performing six operations: (1) ability to generate actions, (2) ability to sense a response from the environment, (3) ability of remembering actions and responses using some form of relational memory, (4) ability to have a goal or target that needs to be achieved, (5) an ability to compare the sensed responses from the environment toward achieving the goal, and (6) ability to somehow use the differences between goal and actual response to guide improvements in the actions.

With this setup, let's look at the steps in the process of learning:

1. The entity generates an action. This action has some impact on the environment.

2. The environment responds to it, and this response is sensed by the entity.

3. The entity then classifies the response as positive, negative, or neutral in the context of the goal it is trying to achieve. It might even generate a numeric

value associated with the goodness of the response to rank it among the positive and negative responses as well.

4. The entity remembers this sequence of events from steps 1–3 in its memory.

5. The entity then uses its memory and differences between perceived responses and target, to create a better action that is more likely to succeed in achieving the goal.

6. Repeat steps 2–5 till the goal is reached.

These steps might look somewhat familiar to the concept of feedback mechanism that we discussed in the earlier part of the chapter. The feedback mechanism does start with the first two steps in this sequence, but then it lacks the steps 3–6 that involve the concept of memory, comparison, and ranking of responses and generation of new actions. Instead, it just responds with a predefined action for the sensed response from the environment. Also, in the machines that employed feedback, the feedback mechanism was explicitly developed by the creator of the machine with anticipation of changes in responses from the environment. The learning mechanism does not make such assumptions. The environment's responses need not be known beforehand and with each new action and response, the memory of the learning entity is being updated.

Most living organisms are going through some form of the learning process from the time they are born. The sophistication and advancements of organs used in each of the steps puts the organisms ahead or behind one another in terms of acquired skills and intelligence. This type of learning is also called real-time learning or online learning or continuous learning. In the case of all living species, this is the only type of learning that is feasible. The learning process is almost as natural as breathing air and digesting food for us, and we don't even realize that it is happening every instant of our conscious life.

Multimodal Learning

Another important concept at play here is around how the sensed response is processed together with the action that was responsible for it. We have many different parts in the body that can generate actions and five different sensory organs as we saw in earlier chapters. The human brain is quite efficient in connecting the responses received from sensory organs like ears and eyes to generating the actions through action-generating organs such as hands and feet. Looking a bit deeper into this process, the actions produced by hands and feet are mechanical in nature that result in a physical movement or change of position of something in the environment, while the signals received from eyes and ears are optical and auditory in nature. These are two very different types or modes of interactions and mixing them together in a meaningful way is not trivial but an extremely difficult task. To illustrate the challenges in this process, let's compare it with Watt's design of governors in steam engine. In this design, the actions of governors were mechanical and produced in response to the mechanical feedback from the speed of engine. There was no camera of any sort watching the operation or any microphone listening to the sounds from the engine. Same applies for all the other feedback machines we saw earlier, from the ancient Chinese chariot to da Vinci's design of self-regulating waterwheel, all the way to the internal combustion engine. Machines that could combine the responses and actions from two different types of media were entirely in a different league. The key missing piece for combining the two interactions from different modes was the existence of an intermediate medium where information from all the other modes can be transformed into or mapped.

The brain inside of all the organisms achieves this through the use of the nervous system. Parts of the nervous system convert all the actions and sensory information into electrical signals. All the action-generating, and sensory organs are connected to the brain through a vast network of nerves and in an average human body, such a network can span a distance as long as 45 miles! With this nervous system in play, all the information that is reaching the brain is now in the same format as electrical signals. The brain can then process this information together in a streamlined manner that actually could have originated from mechanical actions or visual images or touch or sounds or taste and so on. Humans kept on improving the design of the mechanical feedback in machines over thousands of years, but the achievement of multimodal learning kept eluding them. It was only after the mastery on concepts in electromagnetics and invention of computers in the twentieth century that parity was achieved with the technological requirements to emulate organic or human-like learning.

There is one more important aspect in the process of learning that we intentionally ignored in the earlier definition, just to make things a bit easier to understand. However, we can look at it now. It is important to note that the learning entity that we considered in the definition is not always preprogrammed to perform a well-defined action, but rather it is a generic entity, which we can also call as a generic machine, that is equipped with some action-generating artifacts. For example, an arm that can move in certain angles, or a wheel that can move in certain directions with a range of speeds, or an engine that can burn fuel to generate rotation. But in either case, it is not defined what the intended function of the machine is. The arm can be used to push blocks in a construction, or it can also be used to hold another tool to cut a tree. The wheel can be used to drive a cart, or it can be used to build clay pots. Even if the function of the engine is a bit more rigid, the power that is used to drive it can be controlled and manipulated and it can be used for varied applications such as moving a large train from one location to another or moving an elevator up and down. In each case, the operation of the generic machine can vary significantly. None of these applications are known when these machines are created. In order to facilitate multimodal learning, these machines are then provided with a set of sensors and then they are trained using the standard learning steps as outlined earlier.

With such an open-ended start, how would the machine generate new actions at the beginning? Obvious answer would be at random!! However, even when we are generating random numbers, we need to know the range of numbers that is acceptable. In other words, we need to know the lower and upper limits on the numbers to be generated randomly. In the given case, that is provided by the limitations of the generic machine. For example, consider the example of the arm. It would have a limitation on the maximum and minimum angle it can rotate in either direction or limitation on the force it can apply while moving. Thus, using the knowledge of limitations of the machine, it can start producing the actions randomly within its tangible range. Then let's say we have provided this arm with a camera as vision sensor to simulate multimodality, as shown in Figure 6-3.

Figure 6-3. *Robotic arm with vision sensor in the form of a camera*

The camera can observe the motion of the arm. There is a central processing unit on the arm assembly, which we can call the brain of the machine. The brain is generating these actions as electric signals as well as receiving electric signals from the camera in the form of sensory information. This setup takes care of the first two steps outlined in the process of learning. The next steps need to be handled by the brain of the machine. First, we need to define a goal. Let's say the goal is to move an object that is kept in front of the arm to the left by one foot. The brain of the system is able to sense the information from the camera and compare it with this goal and generate a label in the form of good, bad, or neutral. For example, if the arm moves the block in the right direction, it would generate a positive label, if it tries to move it in the opposite direction, it would generate a negative label, and if it does not move at all or moves in a perpendicular direction, it will generate a neutral label. It can even go further to generate a numeric value based on how good or bad the outcome is. Now, we have all the six steps in the online learning paradigm taken care of and we have a generic machine that is ready to start learning.

The brain can start producing random movements of the arm within its limitation and start capturing the labels based on the camera input. This process is also called exploration, where the system is just exploring the action space. If the same process continues for a long time, say a week, the brain will have a huge set of labeled actions recorded in its memory, but the operation of the arm would not improve at all. Now, in order to improve the operation of the arm and get closer to the desired goal of moving the object one foot to the left, we need to introduce the concept of exploitation. Our brain alternates between exploration and exploitation as it continues to record the labeled actions. Exploration means generating random actions without using any of the recorded memory, while exploitation means creating action that it knows would produce a good or positive label or feedback. As the machine starts for the first time, the memory is pretty much empty, and there is nothing to exploit. However, as it continues to explore more, the memory starts to get some actions associated with positive feedback or label. Then it can start to exploit these prerecorded actions and maybe try to make slight tweaks to them to see if the feedback gets better or worse. After such an operation continues for a sufficient amount of time, the arm would reach pretty close to the desired goal of moving the object by one foot to the left. As the machine gets more and more positive samples, it can slowly switch to using exploitation phase only, as there is no need to explore random actions anymore. The goal that we used to train the machine was fairly simple and action space was also quite small and, as a result, such a system can converge on the desired goal fairly quickly. In real life, however, the goals can be quite complex, and the actions spaces are also quite vast, making the learning slow. Exploration and exploitation have their own advantages and disadvantages. Without exploration the machine cannot find new actions and essentially cannot learn. However, without exploitation, it would not produce desired actions in a deterministic manner. So, it is quite essential to maintain a delicate balance between these two operations. These operations are also quite relevant in our lives as well. We are in a strong exploration phase in our childhood, and we learn voraciously; however, we are also extremely unpredictable. However, as we grow old, we start to converge on our learnings and become more and more inclined to use the exploitation phase. Overemphasis on exploitation also makes us rigid and essentially stops any further growth. Great philosophers and religious leaders such as Socrates, Plato, Aristotle, Confucius, and even the Buddha had realized this concept (albeit more along the philosophical context rather than machine learning context) since the ancient times. The teachings and quotes that originated from them such as: "Stay student the entire life," or "never stop to learn" or "question everything" were essentially asking the readers and followers to keep exploring even after you feel you can settle down on just exploiting what has already been learned. There is always something new you can find that will enrich your life even further.

In the context of modern machine learning, the concepts of exploration and exploitation underpin the reinforcement learning paradigm. Reinforcement learning, coupled with deep learning, has proven to be an extremely powerful tool for advancing in the fields of robotics, financial predictions, gaming, healthcare, to name a few. The groundbreaking technology of ChatGPT also uses reinforcement learning to generate text that is both coherent and informative. However, in the case of machines, reinforcement learning is not the only tool that can be leveraged. Other types of learning paradigms can also be utilized that are offline in nature and called batch learning. In the case of batch learning, labeled data containing multiple actions and their corresponding responses are already gathered as a training set and are fed to the learning algorithm at one shot, thereby merging steps 1–2 from online learning for thousands of times. Then the algorithm goes through steps 3–4 as described earlier for all the samples in the training data and completes the learning. At the end of this learning process (also called as training process), the algorithm has created what we call a trained model. This model can now operate in the real world where it can produce actions that would lead to achieving the desired goals. Let's consider an example to illustrate the concept of batch learning. We want to build a face detection system that can detect faces of known individuals. The known individuals can be in thousands or even millions. We have collected photographs of their faces from multiple angles as training data. Now, an average human can remember about 5000 faces. The number can vary based on age, memory capacity, exposure to more or less people, etc. However, that should define the ballpark. We don't expect a human to remember say 100,000 faces. However, we can train a machine learning algorithm using training data containing photographs of faces of a million people using the steps as outlined before. Once the training is complete and we have a trained model, it can start recognizing the faces with accuracy as high as

99%[2] or more, given that the photograph of the person is captured with sufficient resolution and clarity. There are multiple factors that can make this task more challenging such as if the person is wearing sunglasses, or a hat or has grown a beard or moustache or shaved his or her head, etc. The lighting and shadows can also make the task more difficult. However, even after considering all the realistic limitations, such algorithms can do extremely well to the point of being considered as a definitive tool in criminal and legal proceedings. Of course, algorithms need to be supplemented with necessary computation hardware in the form of powerful processors and a lot of memory, but machines are capable of outsmarting humans in such well-defined tasks in a matter of a few hours.

The examples described so far illustrate a high-level overview of how machine learning is executed. Now in the next section of this chapter, we are going to look at some of the leading techniques that are used to train machines to become intelligent.

Machine Learning

The topic of machine learning encompasses all the mathematical, statistical, and computer science developments in building intelligent systems that are capable of learning based on training data. All the techniques studies in this area involve at least two stages of operations in the form of training and testing or application. The application of these techniques can be quite broad and is certainly not restricted to computer science. Typically, this area is considered to be extremely technical and abstract. This is partially true, but most techniques used in enabling the

---

[2] 2018 Facial Recognition Vendor Test (FRVT) conducted by the National Institute of Standards and Technology (NIST); the top 10 algorithms achieved an average accuracy of 99.8%.

machines to learn can be explained purely as concepts, and that is what we are going to do here. We will completely shy away from the mathematics of the theory and look at the concepts that make the machines intelligent by learning like humans.

Linear Regression

Linear regression is one of the simplest techniques in machine learning theory. As you can see in Figure 6-4, the samples are distributed as shown. X axis represents the input to the model and Y axis represents the value to be predicted when the model is ready. These samples can represent the spread of advertising spend relative to business revenue or fuel consumption of a car relative to its speed.

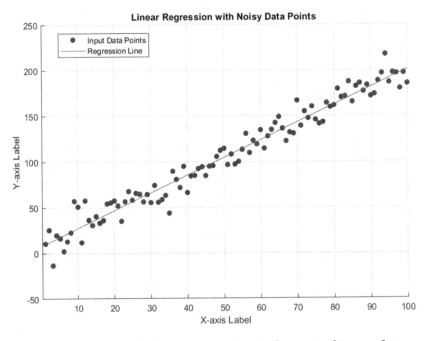

Figure 6-4. *Example of linear regression to fit a set of input data*

These relations typically follow linear trends, and the red line approximates the trend between these entities by a straight line. It is important to note that not all the samples fall exactly on the line, meaning there would be some error in prediction. However, as long as the error is within the acceptable limits, the model is considered good. The essence of the relationship that when value on X-axis increases the value on Y-axis also tends to increase is captured by the straight line. In all the machine learning applications, errors are always present, and zero error situations are extremely rare. Rather when zero error is observed, it can be an indication that something has gone terribly wrong. It is the margin of error that needs to be controlled while training the model.

All the machine learning processes need two sets of data: training data and test data. The training data contains pairs inputs and corresponding outputs or labels. In the context of the preceding examples, each training data would contain pairs of values such as speed as input, fuel consumed as output or advertising spend as input, revenue as output, etc. The test data would contain only the inputs such as speed and advertising spend and using the straight line as predicted by linear regression, we can predict the output in each case in the form of fuel consumed or revenue.

The machine learning techniques, also called algorithms, are sequences of mathematical operations that first operate on the labeled training data to learn patterns of relations between input and output/label and then operate on the test data to predict the outputs. In most real-life situations though, the relation is not linear and as such linear regression algorithm can be quite erroneous, but it helps understand the process of learning.

Decision Trees

One of the oldest and yet one of the most widely used techniques in machine learning is called decision trees. This technique is fundamentally different than linear regression and uses a more humanlike decision-making approach and is more suitable when the problem is to classify given objects into a set of labels rather than predicting a trend. It works by creating a tree-like structure of decisions that can be used to predict the value or type of a target variable. The decision tree is built by starting with the root node, which represents the entire labeled training dataset. Let us consider an example of fruit classification to illustrate the idea. The root node in this case would represent all the fruits. The root node is then split into multiple child nodes by asking a question. In this case, we can ask a question, "What is the color of the fruit?" The answer to that question would create a list of child nodes corresponding to all the colors and all the fruits would be partitioned into these child nodes. No fruit can be part of more than one node. Then each child node can be split further into more child nodes by asking questions about their sizes, shapes, textures, and so on. Each time we ask a new question, we increase the depth of the decision tree. We need to go as deep as needed till we have finally reached a stage where all the child nodes at that depth contain precisely one fruit. The child nodes at this depth are called leaf nodes. Once we reach this stage, the training is complete. Now, we can traverse this tree by asking a sequence of questions and identifying all the fruits in the database. It is important to note that all the paths along the tree leading to detection of the fruit may not be of same length. This explanation was one of the simplest uses of the decision tree. In case of more complex and numerical problems, the splitting process is based on a measure of impurity (called Gini index), which is a measure of how heterogeneous or dissimilar the data is in a particular node. The goal of the splitting process is to create child nodes that are as homogeneous or similar to each other as possible. This is done by choosing the splitting attribute that minimizes the impurity of the child nodes. Once the decision tree is built, it can be used to make predictions by traversing the tree from the root node to a leaf node. The leaf node that is reached will contain the predicted value for the target variable.

Consider another example in the form of a popular iris flower classification problem. There are three types of iris flower species: Setosa, Versicolor, and Virginica. In the experiment, we measured four properties from each of them: petal length, petal width, sepal length, and sepal width. A decision tree model using these measurements would look as shown in Figure 6-5. Note that the depth of the tree is different in different branches, but all the branches end in leaf nodes with a single category of classification. In each block, at the top, we can see the decision rule that is applied to split the data into two sets. The second value shows the number samples in each bucket. Third value states the number of samples in each class in that bucket as ordered list as [# of samples of Setosa, # of samples of Versicolor, # of samples of Virginica]. Fourth value gives the class label. The class label in the non-leaf node can be ignored. From the decision tree, it looks like separating Setosa was quite easy based on just the petal width, but Versicolor and Virginica had a lot of overlap in their properties. Also, you can notice that "sepal width" values are not used in the classification at all, as they may not have sufficient distinguishing properties.

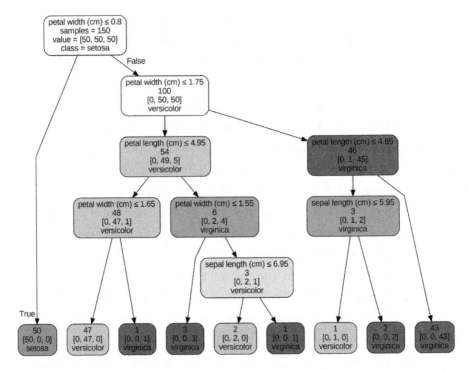

Figure 6-5. *Decision tree classification of three flowers: Setosa, Versicolor, and Virginica*

Decision tree models are relatively easy to understand and interpret, which makes them a good choice for tasks where interpretability is important. They are also relatively efficient to train, which makes them a good choice for large datasets. However, decision tree models can be sensitive to overfitting, which means that they can get too complicated when data contains noise (inaccurate labels). Once we learn all the different machine learning models, one must make an educated guess in choosing the right model for the given problem, as there is no hard and fast rule as to which model would work better in a given situation.

K-Nearest Neighbors

Another example of simple machine learning algorithm is "k-nearest neighbor" or k-NN. This algorithm is similar to making a decision by asking opinions from k-number of your friends. As the name suggests, we need to define the value of "k" for each model. Let's say we define "k" as 5 and the problem is estimating the value of real estate. As part of training data, we have gathered values of a few hundred homes in a given area. When we want to use this data to predict values of new homes that are not in our training set, we follow these steps: (1) We first find the different properties of the house in question, such as number of rooms, square footage, year built, etc. (2) Then we find five homes in our training data that are closest to the house in question based on these properties. These five houses are called five neighbors. (3) Then we estimate the value of the house in question by taking an average of the values of five neighbors as identified in the previous step. This is precisely the process that is used by real estate agents in providing an estimate of your house. Banks use this information before approving a loan. They are all using a classical machine learning of n-nearest neighbors of k-NN.

Artificial Neural Network

Another machine learning model that was invented early on but did not find sufficient success due to its complexity, was called artificial neural network (ANN) or just neural network (NN). This technique of machine learning stems from the way the human brain works, as the name suggests. Figure 6-6 shows a simple schematic of a neural network system.

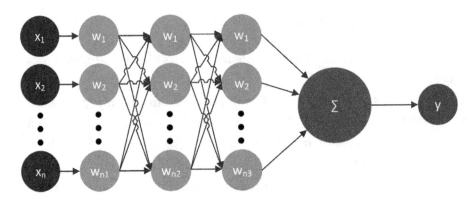

Figure 6-6. *Schematic of a neural network*

Just like the human brain receives its input in the form of electrical signals from all the sensory organs, this computer neural network gets its input $(x_1, x_2, ..., x_n)$ in the form of data (which is carried in the form of electrical signals inside the computer chip). Each circle in the figure represents a digital version of a neuron and each vertical group of neurons represents a layer in the neural network. Each neuron aggregates the signals that are coming as inputs to it and then applies an activation function on the aggregation. The activation function maps values from the input set to the values in output set.[3] The mapping process is similar in all the neurons, but aggregation process differs in each neuron by giving different levels of importance or weight to each of the inputs. Thus, even if the same inputs are given to all the neurons, their outputs are different. Each neuron in each layer of the neural network continues to process the inputs and generate outputs. In the final stage, all the outputs from the previous layer are aggregated to generate final output which can be in the form of an action or prediction. This process is called a feed

---

[3] One of most commonly used activation functions is called ReLU or rectified linear unit. It is mathematically expressed as $f(x) = max(0,x)$. This function essentially zeros out all the negative values and keeps the positive values as is. Another commonly used activation function is called as sigmoid function and has a bit more complex formula as $f(x) = 1 / (1 + e^{x})$.

forward operation of the neural network where the signals keep flowing in forward direction, ultimately converting all the sensory information into generation of an action. After the action is generated, it is compared with the expected outcome that is preprogrammed as per the desired behavior of the neural network. The difference in the generated action and expected outcome is called as error. Now, we need to improve the functioning of the neural network to somehow reduce this error, ideally converging into zero error. In order to do this, a feedback operation is used in the neural network. In the feedback operation, the error from the output layer is fed back sequentially into previous layers. In this step, rather than generating an output in backward direction, the weights that are given by each neuron in aggregating its inputs are adjusted, such that they would do a better job at generating the action in forward direction. This process continues all the way back from the final layer to the input layer. After adjusting the weights for all the neurons in the network, the forward operation is repeated. Typically, after such adjustments, the error is reduced. This new reduced error is then again fed backwards to adjust the aggregation weights of each neuron to further reduce the error. The sequence of forward and backward operations is repeated till the desired level of performance is achieved. Once the neural network is trained to produce the desired level of performance, feedback operation can be discontinued, and only the feed forward operation can be used to generate actions. However, inside the brains of the living organism, feedback operation never stops, as the environmental conditions are always changing, and the organism needs to keep adapting to those changing conditions for the entire duration of life.

The process just outlined represents the most fundamental logical steps that are used in any learning system that is biological as well as computer based. Each logical step is converted into a mathematical form and then it is implemented using python or other suitable programming language. A typical neural network built for the purpose of classifying images or processing language, can require anywhere from few hundreds to thousands to millions of neurons. As such this involves a very high amount of computation, and powerful hardware is needed to successfully implement a neural network-based system. The size of neural network depends on the complexity of the problem that needs to be solved. More complex problems typically require more complex neural networks.

Deep Neural Networks

This architecture of neural network has been in use for more than 70 years. However, since its discovery in the late 1950s, it has gone through many iterations and revisions. As can be seen from the size and complexity of the network, in order to successfully implement such a network, one needs a really powerful computer hardware. In the early days of the discovery of neural networks, the hardware was seriously lagging behind and the architecture could not realize its full potential. By the 1990s, a variety of different algorithms were proposed such as decision trees, support vector machines, etc. that were order of magnitude faster and more accurate than neural networks for solving similar problems. All these algorithms were designed with specific goals in mind and were highly optimized for solving them with limited computer hardware. As a result, neural networks lost their popularity in the field. However, as the computer hardware was getting more and more powerful following Moore's law, by the early 2000s, it came close to realizing the full potential of neural networks. To assist with the central processing unit or CPU in the computer, graphics processing units or GPUs were also reprogrammed to assist in this computation to leapfrog state-of-the-art to the next level. As a result, neural networks started to come back into popularity and became the top contenders in the arsenal of machine learning system. With the added complexity of the system, they are now called deep neural networks or just deep networks. However, at the root, they all have a similar architecture and training mechanism as described earlier. For addressing the more demanding problems in the twenty-first century, some modifications were made into the original structure in the form of adding layers with different type of processing elements. Two of these architectures are most commonly used. One of them is called convolutional neural networks or CNNs and the other one is called recurrent neural networks or RNNs.

CNNs consist of one or more layers of convolutional operators in the network. Convolution is a mathematical operation typically used in signal or image processing. It involves applying a small matrix, known as a kernel or filter, to an input image or data to extract specific features. Imagine you have a grid-like image made up of pixels, where each pixel represents a color or intensity value. The convolutional operation involves sliding a smaller matrix (typically a 3x3 or larger) called the kernel, across this image. At each step, the kernel is placed on top of a small patch of the image, typically starting from the top-left corner. The kernel's values are multiplied elementwise with the corresponding pixel values in the patch. These multiplied values are then summed up. The result of this sum is placed in a new grid, usually called a feature map, or output map, which represents a transformed version of the original image. The kernel is then moved to the next patch, and the process is repeated until the entire image has been covered. The purpose of this operation is to detect and emphasize certain patterns or features present in the image. Different kernels can be used to highlight specific characteristics like edges, corners, or textures in the images. Convolutional operations are especially useful in image recognition tasks because they allow the network to focus on relevant local features while preserving spatial relationships. By repeatedly applying convolutions with different kernels, deep neural networks can learn to recognize complex patterns and objects within images.

RNNs, on the other hand, take a different approach in modifying the structure. In RNNs, the input data is assumed to be sequential data such as time series, speech, or text. Unlike original feedforward operation of the neural networks, which processes data in only the forward direction (except during training), RNNs have connections between their nodes that allow information to flow in cycles (meaning forward as well as backward) even during the prediction phase, enabling them to retain and use information from previous steps in the sequence. Imagine you have a sentence that needs to be analyzed word by word. In a feedforward neural network, each word would be processed independently, but in an RNN, information from previous words is also taken into account as context. At each step of the sequence, the RNN takes in an input (say a word) and processes it along with a hidden state, which acts as the network's memory. The hidden state contains information from previous steps. RNN computes two things: an output and an updated hidden state. The output represents the network's prediction or interpretation based on the current input and the previous hidden state. The hidden state is updated by incorporating information from the current input and the previous hidden state, enabling the network to retain information across different time steps. This process of updating the hidden state and generating an output is repeated for each step of the sequence, allowing the RNN to capture dependencies and patterns in the data over time. The key idea behind an RNN is that it can learn to use its memory to understand context and make predictions based on the sequential nature of the data. For example, in language tasks, an RNN can consider the context of previous words to predict the next word in a sentence. It's important to note that while RNNs can be effective for processing sequential data, they may face technical challenges in capturing long-term dependencies.[4] To address this, variations of RNNs, such as long short-term memory (LSTM) and gated recurrent units (GRU), have been developed with specialized mechanisms to better handle long-range dependencies. However, the biggest improvement in the theory of RNNs was proposed in the form of Transformers.

---

[4] This is due to a problem known as vanishing gradients in recurrent neural networks (RNNs). This refers to a situation where the gradients or the errors used to update the network's weights become very small during the backpropagation process. As a result, the RNN has difficulty learning and capturing long-term dependencies in sequential data.

Transformers introduced a new concept in the architecture of RNNs in the form of self-attention mechanism. This mechanism allows the model to weigh the importance of different parts of an input sequence when making predictions. This architecture goes beyond the use of just neighboring hidden words or tokens from RNNs to identifying the most influential word or token in the entire sentence. This enables the model to capture relationships between all the tokens in the sentence regardless of their positions. To illustrate the impact of this, let's consider the grammatic concepts of subject and predicate in a sentence. Depending on the number of subjects and predicates and their relative positions in the sentence there are many different types of sentences, such as simple sentences, compound sentences, complex sentences, and so on. A simple RNN may not be able to decipher all these nuances, but transformers are not restricted by such limitation and can open up the models to learn any structure thrown at them. However, they need a large amount of training data in order to distinguish all the different types. Another big advantage of transformers is their ability to parallelize the processing. As the model is searching for all the possible relationships in the given sentence of sequence, these searches can be made to run in parallel, unlike RNNs, which only learn in sequential manner. The parallel processing capabilities made the transformer-based models even more attractive as all the newer GPU based high performance compute (HPC) hardware improved the training speed of these models by order of magnitude.

Generative Models

Most of the machine learning algorithms that we have seen so far are primarily used in making predictions or estimates based on some input data. They are more of inference-generating models where output is a classification label or a numeric quantity to infer about the input data. However, machine learning algorithms can also be used in reverse manner to generate data from given labels. These models are called generative models. Although the concept of generative models has come into limelight with the invention of ChaptGPT, these models have been in the works for many decades. Markov Models, or specifically, Hidden Markov Models (HMMs), were some of the first generative AI models. They were used to create sequences of numbers based on predefined probability patterns. As from the perspective of computers, everything can be represented in the form of numbers, these models could also be used to generate text, speech, or even music. The neural networks could also be used as generative models when they are used in encoder–decoder pairs. These models contain a set of encoders that are capable of encoding the input patterns into an abstract numeric form called embeddings. These embeddings can be considered as abstract learnings by the network. Then the decoder part performs a reverse operation generating the input pattern back from the embeddings. Let's consider a simple example to illustrate the encoder–decoder behavior. When we listen to any song, our ears convert those sound patterns into electrical signals and send them to the brain. The neurons in the brain perceive these signals and create their own encodings. This is the abstraction that is continuously happening in our brains. These encodings are then stored into our memory. When we hear a song that we have heard before, the current encoder output in our brain matches with the encodings that are already stored from some time back and it generates a strong positive or negative signal based on whether we like that type of music or hate it. Now, when we try to follow the song in our mind and try to sing it by ourselves, we are trying to decode what we have heard. The decoder part in our brain is connected with our vocal tract and mouth instead of ears, but the decoder is working closely with the encoder as we hear what we are trying to emulate and see how well we are singing. It is important to note that the decoder part is not an automatic inverse of the encoder as we cannot reproduce everything we hear precisely and there needs to be a steep learning curve to match the sounds produced by the decoder with what we had heard through our encoder output from the memory. That is why it takes a lifetime of dedicated practice and learning to play various musical instruments to perfection or to sing the songs with perfect frequencies and tones. However, in the case of machines, learning can happen at an incredibly fast speed with the availability of practically unlimited resources and the encoder–decoder operations can learn to produce perfect inverse in a matter of minutes or hours. Thus, it is far too easy for machines to replicate what they hear or sense, but creating something entirely new and yet good and humanlike is a much harder task.

The encoder–decoder architecture is further refined in the form of generative adversarial networks or (GANs). GANs consist of two separate neural networks, one acts in a generative mode creating sequences of information and the other network acts as a discriminator that compares the output of the first network with real world data and output the differences. These differences are then fed back into the first network, enabling it to learn to produce an output that is close to the one available in the real world. The two networks keep working together, also termed as competing against each other to reduce the error, thereby generating ultra-realistic data.

Even if ChatGPT was one the first widely used language generation model, the history of machine-generated language models goes back to the 1960s, when a simple model called Eliza was created at MIT that simulated conversation by simple matching and substitution techniques. Then in the 1990s, a much-improved version of Eliza was created and was called ALICE. ALICE was a rule-based chatbot that used a library of predefined patterns and rules to generate responses in natural language. ALICE really established itself as a primary precursor to modern chatbots. Then around 2020, OpenAI came up with the first iteration of GPT (generative pretrained transformer) model, and the rest is history. These new transformer models are also called large language models and they are trained on billions of online resources encompassing the entire Wikipedia and many more. Also, it is important to note that they are pretrained models, and they are not actively learning as they are interacting with humans. Just to get an idea of the scale of their operations, let's look at the progression in the number of parameters used in them. The first GPT model that came in 2018 contained about 117 million parameters and truly demonstrated the effectiveness of large-scale transformer models for language generation tasks. The second model or GPT-2 came in around 2019 with over 1.5 billion parameters and showed significant improvements over the first version. It really placed OpenAI at the forefront of the generative AI technology. The third version or GPT-3 that came in 2020 had a mindboggling 175 billion parameters. It showed another huge leap in performance and is also the core model that powers ChatGPT. ChatGPT was a striped-down version of GPT-3 (strictly speaking GPT-3.5) that was released for public consumption. The fourth version of GPT came in around 2023 and contained an over-the-top 170 trillion parameters. GPT-4 is trained on even larger training data making it more accurate in its predictions. The newer architecture is more efficient and scalable making it more cost-effective in generating text in spite of having a much bigger footprint. It is also proven to be capable of generating more creative text formats such as poems, code, scripts, musical pieces, emails, and letters. It can translate between languages with better accuracy, can code more efficiently, and so on.

All these GPT models, albeit showing groundbreaking prowess of AI were restricted to text. There are a whole new set of models that are released around the 2020s such as DALL-E2, Midjourney, etc. that can generate images based on an input prompt. Here is an example image created using a prompt "Elephant wearing superman clothes and riding a bicycle along an ocean in rain," as shown in Figure 6-7.

Figure 6-7. *Elephant wearing superman clothes and riding a bicycle along an ocean in rain*

It shows how far technology has come in understanding the meaning of the prompt and converting it flawlessly into a real rendition of it. This is truly the epitome of the dance of encoder–decoder. These models use a new technology called Stable Diffusion which is based on the concept of GANs instead of the transformer architecture used in case of GPT.

Conclusion

The horizon is just a step away with the progress of machine learning technology these days. Each day sees a new application getting unlocked. You want to create a piece of music for a specific context? Done! You want to architect a bridge to carry certain weight and build using certain materials that can withstand certain environmental conditions? You got it! You want to convert a computer code from one language to another? Yes sir! You want to test the validity of certain assumptions in case of 3D printing an artifact? Just ask!!

In the next chapters, we are going to see where we can go armed with this technology.

CHAPTER 7

Humans in Intelligent Environment: Near Future

For hundreds of thousands of years since the appearance of humans on the face of Earth, humans have been interacting with their environment. The interactions have always been either with living species such as other humans or organisms from other species such as dogs and cats and horses and cows and bulls and so on, as well as trees. Humans have also been interacting with non-living objects such as rocks and soil and water, etc. When humans interact with living species (except probably trees), they always expect some form of active response in return. If we run after a dog, it might run away or may come back charging at us. If we talk with another human, they may talk back or laugh or get scared, etc. Humans are accustomed to such responses. The interaction with trees can be a bit tricky. With no locomotive organs, most of the interaction may seem like one with non-living things, but we can observe that trees grow from seeds, they shed leaves in fall and grow back in spring, they bear flowers and fruits, etc. In general, though, responses to non-living things are always passive in nature. If you hit a rock, you will get hurt, but the rock is not going to actively respond in any way. Also, when left alone, the rock is not going to show any changes to its appearance. Same applies to all the

© Ameet Joshi 2024
A. Joshi, *Artificial Intelligence and Human Evolution*,
https://doi.org/10.1007/978-1-4842-9807-7_7

other non-living things. This is what has been programmed into human brains since eternity. It started to change a little with the introduction of machines into their lives around 7000 years ago as we saw in earlier chapters. These machines were capable of moving things and reacting to human actions in a way that was different than a typical non-living thing, but it was still quite deterministic and predictable. There was no element of surprise or uncertainty in their behavior and humans quickly adapted to work with such machines. For thousands of years from then till the late twentieth century, these interactions were still about the same in spite of vast improvements in the type of machines and their capabilities. The machines were still quite deterministic and predictable. They were still not unpredictable, and no original or intelligent response would be expected from them. The definition of intelligence is deeply rooted in the behavior of living species that come with uncertainty and unpredictability along with logical thinking. As the machine technology reached a point where they started outsmarting humans in complex heuristic calculations (still quite deterministic and predictable) in the form of expert systems, an inevitable comparison with human intelligence started to surface. However, with reference to the notion of human-like intelligence, the machines were still rather dumb.

Since around 5000 years ago or 300 BCE, in ancient Mesopotamia, humans have been using devices such as the abacus that could perform simple mathematical operations. The abacus consisted of a series of rods or wires with beads or stones that could be moved back and forth to perform arithmetic calculations. It provided a simple yet effective method for performing basic mathematical operations such as addition, subtraction, multiplication, and division. Over time, technology evolved into various forms of mechanical calculators, particularly during the Renaissance and Industrial Revolution periods. Notable examples include the Pascaline, invented by Blaise Pascal in the seventeenth century, and the Difference Engine, designed by Charles Babbage in the early nineteenth century. These early mechanical calculators paved the way for the development of more sophisticated and advanced calculating devices, leading to the invention of modern electronic calculators in the twentieth century. These gadgets could perform non-trivial and complex arithmetic operations such as finding square roots and factorials and exponents with very large numbers in a matter of seconds; something that most humans cannot even dream of. Electronic calculators really proved a point that machines had gone way beyond human capabilities as far as numerical calculations are concerned. The twentieth century also saw the emergence of computers, which were significantly more advanced devices that could perform logical operations on top of basic arithmetic calculations. The processing power of computers has grown exponentially every year since their inception and by the early twenty-first century, some of the most powerful supercomputers have already surpassed the human brain in terms of raw processing power and memory capacity.

Carbon and Silicon

It is a good place to digress a little to understand the difference between the computation happening in the human brain against the computation happening inside of computers. The differences start at the atomic level, where humans and all the living organisms for that matter are based on Carbon, while all the computers and electronics are based on Silicon. As we can see in the periodic table (Figure 7-1), Carbon and Silicon occupy a very similar position. They have different (6 and 14, respectively) numbers of protons and electrons in their atoms, but both have exactly 4 electrons in the outermost orbit, which can contain anywhere from 1 to 8 electrons.

This makes both these elements the most versatile in terms of capability of bonding with other elements. The outrageously long molecules such as our DNA can only form through the chain of billions[1] of carbon atoms.

Periodic table of the elements

Legend:
- Alkali metals
- Alkaline-earth metals
- Transition metals
- Other metals
- Other nonmetals
- Halogens
- Noble gases
- Rare-earth elements (21, 39, 57–71) and lanthanoid elements (57–71 only)
- Actinoid elements

period	group 1*	2	3	4	5	6	7	8	9	10	11	12	13	14	15	16	17	18
1	1 H																	2 He
2	3 Li	4 Be											5 B	6 C	7 N	8 O	9 F	10 Ne
3	11 Na	12 Mg											13 Al	14 Si	15 P	16 S	17 Cl	18 Ar
4	19 K	20 Ca	21 Sc	22 Ti	23 V	24 Cr	25 Mn	26 Fe	27 Co	28 Ni	29 Cu	30 Zn	31 Ga	32 Ge	33 As	34 Se	35 Br	36 Kr
5	37 Rb	38 Sr	39 Y	40 Zr	41 Nb	42 Mo	43 Tc	44 Ru	45 Rh	46 Pd	47 Ag	48 Cd	49 In	50 Sn	51 Sb	52 Te	53 I	54 Xe
6	55 Cs	56 Ba	57 La	72 Hf	73 Ta	74 W	75 Re	76 Os	77 Ir	78 Pt	79 Au	80 Hg	81 Tl	82 Pb	83 Bi	84 Po	85 At	86 Rn
7	87 Fr	88 Ra	89 Ac	104 Rf	105 Db	106 Sg	107 Bh	108 Hs	109 Mt	110 Ds	111 Rg	112 Cn	113 Nh	114 Fl	115 Mc	116 Lv	117 Ts	118 Og

lanthanoid series 6

58 Ce	59 Pr	60 Nd	61 Pm	62 Sm	63 Eu	64 Gd	65 Tb	66 Dy	67 Ho	68 Er	69 Tm	70 Yb	71 Lu

actinoid series 7

90 Th	91 Pa	92 U	93 Np	94 Pu	95 Am	96 Cm	97 Bk	98 Cf	99 Es	100 Fm	101 Md	102 No	103 Lr

Figure 7-1. Periodic table showing position of Carbon and Silicon

All the electronic chips are made out of wafers of Silicon. Silicon, on top of having exactly 4 electrons in the outermost orbit, also has special chemical and electrical properties that make it a semiconductor. Semiconductor elements are elements that have a conductivity between that of an insulator and a conductor. This is a crucial property that makes

---

[1] In human DNA, there are approximately 3 billion base pairs, and each base pair contains a sugar molecule along with one of four nitrogenous bases (adenine, thymine, cytosine, or guanine). This means there are 3 billion sugar molecules in total. Each sugar molecule contains five Carbon atoms, while nitrogenous bases do not contain carbon atoms. Therefore, the estimated number of carbon atoms in human DNA would be around: 3 billion (base pairs) x 2 (strands) x 5 (Carbon atoms per sugar molecule) = 30 billion carbon atoms.

Silicon an ideal material for building electronic chips. Germanium and some compounds of Gallium and Indium can also act as semiconductors, but those elements are much more complex and heavier than Silicon and not good candidates for chip manufacturing. Surprisingly, in spite of being lighter than Silicon, Carbon[2] is not a semiconductor and hence cannot be used in chip manufacturing. Semiconductors can be easily controlled to either be a good conductor of electricity or block the electricity entirely as insulators. This is a fundamental property that enables the creation of logical units in the chips that lay the foundation for all the complex processing that can be programmed into them. The first semiconductor chip, also known as an integrated circuit (IC), was produced by Jack Kilby at Texas Instruments in 1958. Kilby's invention marked a significant milestone in the field of electronics and is considered to be the birth of the microchip. Kilby's chip was made using Germanium, not Silicon, as Silicon technology was still in its early stages at that time. The chip consisted of a wafer of Germanium with several components, such as transistors and resistors, interconnected on its surface. It was the first demonstration of integrating multiple electronic components into a single substrate. However, shortly after Kilby's work, Robert Noyce, co-founder of Intel Corporation, independently developed a practical Silicon-based integrated circuit in 1959. Noyce's Silicon-based IC became the basis for

---

[2] Carbon is not a semiconductor because it has a large band gap. A band gap is the energy difference between the valence band and the conduction band. The valence band is the band of energy levels that electrons can occupy in a solid. The conduction band is the band of energy levels that electrons can occupy when they are excited. In a semiconductor, the band gap is small enough that electrons can be excited from the valence band to the conduction band by absorbing a small amount of energy. This allows electrons to flow freely in the conduction band, which is what allows semiconductors to conduct electricity. Carbon, on the other hand, has a large band gap of about 5.4 eV. This means that it requires a lot of energy to excite electrons from the valence band to the conduction band. This makes it difficult for carbon to conduct electricity, and therefore carbon is not considered a semiconductor.

the modern semiconductor industry due to the inherent advantages of Silicon as a material with better thermal properties and higher reliability. Kilby and Noyce opened the door to the revolution that followed with creation of Silicon Valley and were rightfully awarded the Nobel prize in the year 2000. The process of manufacturing the microchip is called photolithography. It helps to create tiny, intricate patterns on the surface of a silicon wafer, which is the base material for the chips.

Manufacturing of Integrated Chips

Imagine you have a plain piece of glass, and you want to draw a very detailed picture on it. However, you don't have a brush that is tiny enough to paint the picture directly. So, what you do is place a special stencil called a mask on top of the glass. This mask has all the intricate patterns that you want to transfer onto the glass. Then, you shine a bright light through the mask onto the glass. The light passes through the clear areas of the mask and exposes the glass, but it gets blocked by the opaque parts of the mask. This way, the light creates a pattern of exposed areas on the glass, following the patterns on the mask. Now, you can apply a chemical called an etchant to the exposed areas. This etchant can remove or alter the properties of the glass, depending on the specific process. After the etching process, you remove the mask, and you're left with the intricate pattern transferred onto the glass. In the case of photolithography, the "glass" is actually a Silicon wafer, and the "mask" is a high-resolution photograph, or a pattern created using computer-aided design. The light used is usually ultraviolet light, which can create very fine details. The patterns created through this process are crucial for building the complex circuits that make up electronic devices. By repeating the photolithography process multiple times with different masks and chemical treatments, manufacturers can build up multiple layers of patterns on the silicon wafer, creating the intricate circuitry needed for modern electronic devices.

Before going into the details of logic operations, let's quickly review how logic is represented in ICs. Consider we have two input signals, A and B. Each signal can have a value or *True* or *False*. So, we have a total of four possible combinations of inputs between A and B as shown in Table 7-1. When we apply AND operation on them, the output is *True* only if both the inputs are *True*, otherwise it is *False*. To apply this logic in a real situation, consider a couple of friends, Adam and Barry. If Adam and Barry both want to play chess, they play chess, if either of them is not interested then they don't play chess. The above-mentioned AND operation precisely mimics this situation. Now, consider another case. If either Adam or Barry wants to go for movies, they both go for it. Only when both of them are not interested in going to the movies, they don't. This operation is mimicked using the OR operation on an IC. The output of OR operation is *True* when either of the inputs A or B is *True*, and *False* otherwise. These two operations form the foundation for implementing any type of logic on ICs. We can add NOT operation on top of them, or combine them with additional inputs and just add a series of such operations back-to-back to simulate any real life situation involving logic.

Table 7-1. *Logic operations on the IC*

Case #	A	B	A AND B	A OR B
1	True	True	True	True
2	True	False	False	True
3	False	True	False	True
4	False	False	False	False

These logic operations are implemented in the IC using a module called "Transistor." A single transistor represents a unit of computation that can perform a basic logical operation such as "AND" or "OR" in an IC. With an appropriate combination of these operations (albeit in millions or even billions and more) one can create any arbitrary heuristic process one can think of. Since the introduction of the first IC, the number of transistors on the chip has nearly doubled every two years, driving the exponential growth in computation power. This empirical observation is attributed to Gordon Moore (co-founder of Intel) and is widely known as Moore's law. There is a finite limit to this exponential growth bounded by the size of the atoms of Silicon and wavelength of the ultraviolet light. With the increase in the number of transistors, one needs to make the circuits more and more miniaturized to keep the size of the IC within reasonable limits, thereby increasing the complexity of the process of manufacturing. As of year 2020, we have reached the process where the smallest transistor is created in a space of about 5 nanometers. Theoretical limit for the process is about 1nm bound by the wavelength of the ultraviolet light that is used in manufacturing. Moore's law will theoretically end at that point.

These Silicon-based machines can perform numeric and logical operations at an incredibly fast speed and accuracy. Thus, if a solution to a problem can be outlined with finite lines of logical statements (it can be in millions or more), computers can be programmed to solve it in a finite time. The time can further be reduced arbitrarily by increasing the processing power and speed of the machines. Despite machines mastering mathematics as well as logic, they still could not pass the quintessential Turing test for a long time to be intelligent in a human-like manner. The critical remaining piece was interaction in natural language. Computers were still not capable of understanding or interpreting natural language and responding to it by the end of twentieth century.

Extrapolation and Human-like Intelligence

Along with being logical, deriving insights from past experiences and being vague at times are also key ingredients in human intelligence. In human conversation, not every sentence is precise and has a clear binary or explicit logical meaning. There are expressions that can mean multiple things, that are unclear, or sometimes even downright contradictory.

Computers typically cannot produce such responses, rather they are meticulously programmed not to produce such responses. So, what if we change our objective in training the computer algorithms and make them produce such vague or multifaceted or even contradictory remarks? Would they start showing human-like intelligence? There is a thin line between producing random and meaningless responses and responses that draw from past experience, have some logic and still be vague or contradictory. At this point, we should look into another aspect of learning: extrapolation. In technical terms, extrapolation is a way of predicting values that are outside the range of training data. A simple example of extrapolation would be predicting the values of a stock price or predicting a weather forecast. The training data for the problem of stock price prediction or weather forecast is always from the past and the predictions are always for the future where the model invariably encounters something new. However, this being a strictly numerical problem and numerical values are rather too well defined and constrained from top and bottom, the predictions in the extrapolation mode are still relatively easy. Still, if the model is asked to predict the future stock prices of Microsoft and is trained only on the historical prices, the predictions are likely going to stay in the range bounded between the lowest and highest historical values. In the same way, the weather forecast for New York city would likely stay withing the bounds of the extremes that have been encountered in the past. The model would most likely fail to predict accurate weather when an entirely new type of weather pattern emerges. In same way, if something entirely new happens in the market, especially with respect to the competitive landscape, the model is not going to be able to understand its impact, and as a result, not able to predict the stock values of Microsoft that are likely going to be quite different from the historical values. Furthermore, not all the problems that need extrapolation are numerical. Consider an example of a computer model that is trained to classify animals into two categories: dogs and cats. All the training data that was given to the algorithm only consisted of pictures of different breeds of cats and dogs. If this model is now presented with a picture of a chimpanzee, the outcome of the model can be quite unpredictable. In this case, when the model is trying to classify the chimpanzee into one of the only two categories that it is trained in and that do not contain a chimpanzee would be called as extrapolating. Multiple things can happen in this case. The model can either just classify the chimpanzee as a dog or a cat. It is also possible that the model would just fail, and the program may crash. None of these responses are "intelligent" or "smart" from the perspective of human intelligence. A human in such a situation would most likely come up with an answer "I don't know!" Right from birth, humans or all the living organisms for that matter are always facing situations that are new in some ways and not encountered earlier, and they are always expected to respond to these situations without having an option to crash. In computer terms, living organisms are almost always doing extrapolation and are rather quite adept at doing it. The examples of stock prediction or animal classification were oversimplistic ones and as such examples like these are rarely observed in real life. The extrapolation in real life is quite nuanced and comes with elaborate context. For example, after getting a driver's license, when one starts driving, each time they drive on a new road, a new traffic situation is encountered. Each new neighborhood would present different combinations of intersections, roundabouts, pedestrians, freeway exits and entrance ramps, and so on. None of these situations were presented during the learning sessions in exactly the same way and each person in that situation is still figuring out how to respond to these situations and most of the time, they are doing a respectable and accurate job. However, training machines to do well in such nuanced extrapolation environment is a much harder task and it took a relatively long time to develop the necessary technology. We looked at the concepts of exploration and exploitation in the context of human learning in the previous chapter and there is definitely some perceived similarity of extrapolation with them. As a matter of fact, both the operations of exploration and exploitation are needed when operating in extrapolation. Whenever a new situation is presented, one needs to find its closeness or proximity to the situations that were encountered in the past, find out what were the responses that received good feedback and then create a response that is a combination of them to generate a new action. Also, to infuse exploration aspects in the mix to enable new learning, there must be some randomness infused in the response as well. Thanks to the innovations in deep learning and transformer techniques, we have figured out ways to embed such behavior in the learning algorithms and programs, like Open-AI's ChatGPT are great examples of bringing such learning to action.

In spite of making huge strides in the field of artificial intelligence in the early twenty-first century till 2020, the benefits of the technology were quite limited to the people in the know. The people who understood the concepts of computer science and advanced mathematics were the only ones who could really appreciate and leverage them. The only languages these intelligent machines spoke were coding languages that most of the world was oblivious to. ChatGPT changed the game for the first time, when it enabled natural language interaction with the AI and opened it to the entire world for free. For the first time in the history of machine intelligence, ordinary people, people with no prior knowledge about technicalities of machine learning and computer science could interact with AI in their natural language and get results that they can understand. ChatGPT proved to be the most viral application in the history of computers, reaching over one million users in under five days. To bring that into perspective, it took Netflix three and half years to reach that landmark, while Spotify took ten months to climb there, the biggest social network Facebook took nine months to achieve the same, and Twitter needed seven months, while Instagram ran past it in two months. ChatGPT is not a social networking platform, nor is it a music streaming app. It does something that no other app has even attempted before. One million user-base was only a preface to the story of ChatGPT and it went on to reach 10 million in just 40 days, 100 million in 60 days and on its way to reach over a billion users as of the time of writing this book. This unprecedented success of ChatGPT is not necessarily all about ChatGPT itself, but it is about the value that AI can add into the human life from all the different backgrounds. ChatGPT provided a simple outlet where anyone in the world could access the cutting-edge capabilities of AI. Each time a new user is sampling ChatGPT, a new chapter is being written in the story of the human race. Each new question that is being asked of ChatGPT is paving the way for a revolution in some new field. The life-altering changes that industrial revolution brought in a matter of a couple hundred years, the modern AI is positioned to offer in a matter of a few decades, if not sooner. The current generation is going to face a whole new tomorrow when they are going to graduate, a whole new job market, a whole new set of challenges, whole new ways to interact with each other and whole new ways to get entertained.

Adapting to Intelligent Machines

Machine intelligence made leaps and bounds progress starting with the twenty-first century and the gap between human intelligence and machine intelligence started to get blurred to the point where machine intelligence was almost impossible to distinguish from human intelligence. This change marks a milestone in human history as big as the milestone that was marked by the creation of the first machine about 7000 years ago. Machines helped humans become more powerful to achieve things that they could think of but could not implement by themselves. It made their lives easier, safer, and richer. It also meant humans had to adapt to live with machines to really benefit from them, and humans did that quite effectively over the course of the past thousands of years. The period of 7000 years may seem like a long period of time, but when we look at it from the perspective of human history spanning hundreds of thousands of years, when humans were living with mere simple tools, it would appear a very short time.

Starting in the twenty-first century, with the appearance of machines that are as intelligent as humans, it is going to take human interaction with non-living things to another level. The relatively fast pace of technological evolution in development of simple machines over the past thousands of years is now going to look excruciatingly long compared to only decades of time it is likely to take for the AI powered machines to take over. With just a blink of an eye, it's going to be a very different world. The advent of machines helped humanity reach new peaks, but at the individual level, it was quite an uphill battle as the bulk of the people lost their traditional jobs and had to adapt to the new era of living with machines. The process was accompanied by multiple sociopolitical revolts until the civilizations across the planet settled into the new norms.

The current adaptation that is going to be needed is not only paced an order of magnitude faster but also has deeper-reaching roots and is going to fundamentally alter some of the concepts that humans have been carrying with them in their genes since their origin on the face of Earth. Let's look at them one by one.

SuperAI

At this point, it would be useful to make a distinction between some of the different applications of machine intelligence. As we saw in previous chapters, predicting real estate prices or weather forecasts and stock prices are considered as applications of machine intelligence or AI. However, in the light of new human-like super intelligence in the form of generative AI applications, these earlier examples of intelligence fade away. So, we will call the traditional machine intelligence mostly from the twentieth century as just "intelligence" or AI and the new machine intelligence, starting in the twenty-first century that is powered by deep learning, as super intelligence or SuperAI.

At the time of writing this book (around mid-2023), the extent of the SuperAI machines is rather limited, primarily restricted to mining the Internet or world wide web and making it accessible in the form of an interactive conversation bot that can answer any question in natural language or generating a new piece of information in the form computer code, small literary artifacts, images, poems, and so on. It has already shown promise in creating artifacts that would otherwise take hours, if not days and weeks, to create by skilled humans. Most of the interaction as of now is limited to written text only, but it can change quite rapidly. With the availability of voice assistants such as Amazon's Alexa, Google's own Assistant, or Apple's Siri, this technology can be easily adapted to voice-based interaction, but that has not happened yet at the time of writing this book. In a shockingly short duration of time, these algorithms have proven that machine intelligence is ready to take over an extraordinary number of jobs from humans. There are caveats with respect to how fast SuperAI can achieve it and must be looked upon on a case-by-case basis. In the next part of this chapter, we are going to take a comprehensive look at each of the broad categories of jobs that humans do in most developed and developing countries and how SuperAI is going to affect those.

Impact of SuperAI on Job Market

Agriculture

Let's start alphabetically with agriculture. Agriculture is already getting benefits from AI machines in the form of satellite and drone imagery to monitor crop health, predicting the appropriate use of fertilizers and pesticides. This has made a revolutionary improvement in farming and agriculture as a whole by reducing the dependence on unpredictable natural patterns and calamities. These predictions, followed with mitigations, have made agriculture a lot more resilient and efficient. Robots are used for handling repetitive tasks such as planting, weeding, harvesting, etc., thereby reducing human dependence on farm animals and making these tasks scalable and fast. SuperAI applied in agriculture (let's call is SuperAgriAI) can tie up the remaining loose ends with these operations where humans are involved, such as analyzing the data and recommending and then deciding which action to take. Managing communication and business operations can also be automated with SuperAgriAI. It would further improve the precision in agricultural techniques, optimizing the crop yields and reducing other costs. Integrating and automating the communication between all these separate entities would leave precious little work for the farmers to the point of making their existence almost redundant. Imagine a scenario where robots are working in the field, doing all the farming operations as directed by the central computer, a part of SuperAgriAI. This SuperAgriAI has access to satellite images as well as control of a group of drones to obtain all the details about the health of the crop and surrounding environment. The SuperAgriAI knows all the details about the soil and weather patterns for the given area, chooses the best category of crops to plant, and employs the robots that are doing all the necessary operations such as watering and application of pesticides, etc. at precisely the right time in precisely the right amount. All the robots are also equipped with sensors to detect any malfunction, and SuperAgriAI can work out the necessary maintenance schedules for them to keep them operating optimally. SuperAgriAI communicates with the other businesses that provide all the necessary goods and makes the financial transactions as needed, ultimately keeping the entire platform of food production operating flawlessly end to end. In the same way, this SuperAgriAI can control livestock management with robots that are specialized for that work. With all the modern handling of farm animals, we are treating them as commodities anyway, so their complete management through SuperAgriAI would be a natural culmination of modernization. The same concept can be applied to fishing. SuperAgriAI can control fully automated vessels carrying all the fishing equipment along with various types of sensing equipment such as radar and sonar for locating the ideal place to fish. SuperAgriAI can apply the fishing nets with pinpoint accuracy to catch only the desired amounts of fish without inflicting any collateral damage. With continuous feed from the weather system, it would also choose the best times to send the vessel. Being operated strictly by robots, there is no need to account for time of the day or track the holiday schedule or even plan for any meal breaks. Fisherman's days are numbered as well. This still leaves people like horticulturalists or fishery biologists who are doing research with crops and plants and fishes, etc. In the near term, their jobs seem to be fairly secure. It is quite an open-ended job that tracks the current operations of agriculture and fishing, essentially getting all the inputs from SuperAgriAI and trying to find ways to do better. Another related job area is that of forestry. Foresters are professionals who manage forests. They use their knowledge of forest ecology, wildlife biology, and forestry practices to plan and implement projects that protect, secure, and manage forests. A majority of their field work can be taken up by robots and SuperAgriAI, leaving a small chunk of jobs that strictly focus on research. Landscapers use their expertise in plants to create innovative and beautiful landscapes. Considering SuperAgriAI would have all the necessary knowledge about the plants with robots to do the manual work, landscapers would also need to look elsewhere. However, it would still be an open question as to who would create and manage this SuperAgriAI. Currently the technology required to create SuperAgriAI is only available with top tech companies in Silicon Valley, but the subject matter expertise lies with other agricultural companies. Whoever takes the first step is likely going to get a head start.

Business

Business is an extremely broad category that likely provides the most jobs in the modern world. In its overarching sense, all the categories can be grouped under business, but we will look at it from a narrower interpretation to separate different fields for dedicated evaluation.

From a financial perspective, SuperAI in business (let's call it SuperBizAI) is going to provide a plethora of tools to manage the accounting and payrolls type functions. Let's look at the top job profiles in this area. Accountants are responsible for creating and maintaining the financial records of a company. They also prepare financial statements and track income and expenses. There can be accounts specialized in subareas such as audit, tax, and financial planning. The entire profile of accounts can be fully automated with SuperBizAI with very little risk. It is only matter of current accounting applications to mature to the level that big organizations can trust them with full responsibility. Again, the tech giants have a huge head start in this area as well. Other than accountants, there are roles such as auditors, payroll clerks, tax preparers, etc. and these roles are also on the path to becoming extinct with SuperBizAI ready to replace.

Banking

Banking is another prominent area in the business domain. Key operations that are involved in banking include customer service, financial and investment consulting and transactions, credit and risk management. ATMs or automated teller machines were the first entry of machines into this area, and they have already reduced the workforce required for managing basic "teller" type of activities. SuperBizAI can take it a step further by providing automated experiences for human-like interactive customer service. This has been a huge bottleneck in recent years where the pre-programmed automated bots are extremely limited in functionality and can annoy customers more than helping them. ChatGPT like interfaces can change the landscape drastically. The algorithms provided by SuperBizAI can tackle even more advanced banking activities such as loan and mortgage consulting, investment consulting. SuperBizAI can easily personalize all the banking requirements for each individual customer by providing in-home interactive experience. Once realized, it would eliminate the need to have local branches for any bank (only ATMs would need to be placed in strategic locations). The central office can handle the worldwide operations in a fully automated manner. This entire area is now on the verge of extinction from the perspective of job prospects. Along with banking and financial businesses, the other consulting activities are also on track to become obsolete as SuperBizAI with its vast repertoire of knowledge of laws and rules and past events can provide personalized consultations, analyses better than any individual.

Hierarchy of Roles and People Management

In the business domain, there are job profiles based on different industrial sectors and applications, but there are also different job profiles based on the hierarchy of the jobs in a given company. The tree of hierarchy starts from the CEO or president at the top and then goes down to upper-level management to mid- to lower-level management all the way to individual contributors. Each company may have a different relative structure, but the roles of individual contributors and managers are industry standard along with top executives. So far, we have looked at mostly the individual contributor type roles in businesses, where SuperBizAI can play a role, but as we go up the chain of management, there are additional job responsibilities. Mid-level managers certainly need subject matter expertise, but on top of that, it is also important for these roles to have good people management skills. This area has been studied heavily in recent years and there are books that discuss these aspects philosophically as well as based on surveys. One particular book I would like to mention in this context is *First break all the rules: What the world's greatest managers do differently* by Marcus Buckingham and Curt Coffman. The findings in this book are based on one of the largest surveys ever conducted in this area that included real people managers from a variety of different industries. The survey included over 80,000 managers and it was conducted by Gallup. One of the key findings from the survey was that there exists no constant set of golden rules that each manager needs to follow, rather there were entirely contradictory behaviors found in managers that were all extremely successful at their job. Five aspects came out at the top that most successful managers do well: focus on strengths, set clear expectations, create positive work environment, delegate effectively, and hold employees accountable. The job of people management is something that is never associated with machines so far and it was quite appropriate given the limitations of the machines in human-like interactions. However, with SuperBizAI lurking on the horizon, things are quite different now. With such extensive studies stating the top targets for mid-level managers, SuperBizAI can tackle this task and likely tackle it like no human can ever dream to. When SuperBizAI takes the role of a mid-level manager, the issue of subject matter expertise is automatically taken care of. Another and even more critical advantage is that SuperBizAI does not need to be paid like typical managers. It would be purchased as a service by the entire organization with annual fees and those fees would be substantially less than what the organization would have to pay for the management. Now, we need to look at all the tools that SuperBizAI can have at its disposal for people management. It can easily be empowered with continuous monitoring of the employees during work hours. This is something that a human can never have due to obvious invasion of privacy, but machines are not measured with the same standard and monitoring the activities of employees during their work hours by machines is generally not considered as overly intrusive. Already, in many places, there are surveillance cameras tracking employee activities. Some privacy issues can be involved in such monitoring in the new hybrid work culture if the work is happening at home, there can be innovative ways to circumvent those. With full information about the employees' active work and the results that they have produced, SuperBizAI can make an impartial and unbiased evaluation of the employees quite likely far more accurately than any human. Historical data around the working hours and results produced from employees can easily be used to identify their strengths and weaknesses, enabling the SuperBizAI to set the right expectations for the future and also delegate the work optimally. The monitoring would also help in identifying any roadblocks encountered by the employees in real time and immediate action could be taken to mitigate them. Identification of strengths and weaknesses coupled with aspirations of the employees around their future career can help provide customized recommendations to them on what can be improved/changed. Even the most capable and successful managers are still humans and are prone to unconscious biases that are inevitable when dealing with all the reporting employees, meaning they cannot treat all the employees equally or fairly. However, this is not true for machines. With the possible lack of any type of unconscious biases, they can provide perfectly objective treatment to all the employees. This inherent fairness in the operation of machines would automatically cultivate a strong positive atmosphere in the team. One need not do anything to please their manager. One need not worry about truly expressing oneself about creating any conflict. All one needs to do is complete the job that is given, and there would be assurance that they will be rewarded solely based on that. Furthermore, SuperBizAI can operate in parallel and can be omnipresent, thereby making itself available to all the reporting employees all the time. This would inculcate a strong sense of caring in the minds of the employees and also enable even more flexible working hours.

Typically, all the mid-level managers reach this level by first being in the role of individual contributors. So, as they are leading a team, the company as such is losing that particular individual resource from the perspective of the work they can do as an individual contributor. Thus, completely eliminating the mid-level management through extensive use of SuperBizAI can provide a win-win situation for all the companies. The experienced individuals can continue to grow in their career and be more impactful in the decision-making process and leading the projects, but people management aspect can be left for SuperBizAI. There are other more social aspects to the people management such as coaching and mentoring. However, they can be handled independently with either SuperBizAI or some dedicated personnel as needed that are either subject matter experts or experts in psychology. Another task that also falls in the buckets of mid-level management is planning. Planning for short- and long-term goals as per the direction the overall organization is moving toward. In this task, SuperBizAI can potentially generate a list of options that can get augmented and voted on by the senior people in the team to finalize the plan.

Building mature software that can take up roles of mid-level management seems like a nascent area for new startups at the time of writing this book. Once again, all the tech giants can have a head start in this matter as they already have a multitude of existing positions that they can replace and a deep know-how in software development. It is important to note that there can be a strong potential pushback against this in the early stages from all the existing mid-level managers that each organization would have to work out in their own way. This can actually help the upcoming startups to take a lead.

As the application matures, it can work its way up the chain in the management, leading to top-level management as well. However, replacing the personnel at the top is an entirely new concept and can be further along the road. Chief executives of an organization are the primary decision-makers with respect to the direction the entire organization is going to move forward and as such is not a very directly replaceable attribute for the SuperBizAI. However, there are still some pros to consider. For example, being true to company mission and vision all the time may not be possible for humans. Even these executives are humans, and they are prone to human weaknesses, such as (1) overconfidence: Top executives might become overly confident in their decisions and strategies, leading them to underestimate risks and make hasty choices without properly assessing potential drawbacks. (2) Confirmation Bias: Top executives may seek out information that confirms their existing beliefs and opinions, disregarding contradictory data. This bias can lead to poor decision-making and missed opportunities. (3) Lack of Adaptability: In a rapidly changing business environment, executives who are resistant to change or new ideas might find it challenging to adapt their strategies to evolving circumstances. (4) Overlooking Ethics and Compliance: A lack of attention to ethical considerations and regulatory compliance can lead to reputational damage, legal issues, and decreased stakeholder trust. (5) Difficulty with Feedback: Top executives who are not open to constructive criticism and feedback may miss opportunities for improvement and fail to address organizational issues. (6) Poor Communication Skills: Ineffective communication can hinder top executives' ability to convey their vision, motivate employees, and align the organization toward common goals. (7) Burnout and Stress: Top executives often face high levels of stress and pressure, and if not managed properly, this can lead to burnout and impact decision-making. (8) Hubris: An excessive ego or sense of entitlement can lead to decisions that prioritize personal interests over the well-being of the organization and its stakeholders. (9) Groupthink: Top executives who surround themselves with like-minded individuals and discourage conflicting opinions can create an environment of groupthink, which limits creativity and innovation. (10) Failure to Empower Employees: Top executives who do not empower and support their employees may stifle initiative and prevent the organization from harnessing its full potential.

Each and every aspect of these weaknesses can be categorically avoided with proper configuration of the SuperBizAI that is going to replace the top executives in an organization. On top of that, all the most senior people in the company can participate in voting on the key decisions whose options would be curated by SuperBizAI. The voting can also be opened to all the employees of the company or even to all the stockholders of the company with adaptive weighting of the votes to truly create a transparent atmosphere and give a real sense of involvement to all the employees and stockholders in the company operations.

Especially for the large publicly traded companies, from the stockholders' perspective, a CEO or any top executive should be entirely focused on the betterment and growth of the company all the time even during their off-work activities, which may not be humanly possible. However, SuperBizAI does not have such limitations. The process of replacing the management chain can begin at the bottom and as the process matures, it can grow all the way till all the managers are replaced with SuperBizAI all the way to the CEO. All the humans would serve as individual contributors, with appropriate levels of seniority. Humans can now relinquish the aspects that contradict with their personality and stay free to live their non-work life the way they want in a true sense. Even the most senior personnel need not worry about the public image, as that part would be left for SuperBizAI, and which can provide an exemplary performance in that role.

Impact on Jobs with Varied Skill Levels

Moving to specific roles of individual contributors, they can vary from less skilled and labor-intensive factory workers to highly skilled workers such as actuaries, pharmacists, lawyers, software engineers, scientists, and doctors. The numbers of most labor-intensive jobs in engineering and construction industry have already been reduced with heavy use of robotic machines. Imagine a construction site for building a high riser. In the early days of automation, hundreds of individuals would be required to handle numerous labor-intensive jobs at such sites. However, with the availability of heavy machinery such as cranes, excavators, and material handling robots, that number reduced by order of magnitude. Robotic systems can help prefabricate most of the components off-site and assemble them with precision on-site. Addition of computer-assisted 3D printing can create complex architectural elements and building components with far more accuracy with further reduction in semi-skilled workers needed for crafting intricate designs. Use of specialized robotic tools can handle entire demolition jobs, minimizing the workforce and reducing the risk to human workers as well as increasing efficiency. Drones and robotic surveying equipment can collect accurate data on construction sites, eliminating the need for manual surveys and further improving accuracy. Robotic bricklaying systems and concrete pouring pumps can accelerate the other manual steps in construction, further reducing the workforce needed. With remote and virtual reality-assisted operations, humans need not even be physically present at the work sites but can stay at their homes or other favorite places and can handle or monitor all the machinery using virtual reality devices. Most of these devices are still classified as traditional AI. With SuperAI powered robots, the requirement for human presence would reduce further, but will likely still not completely vanish in the near term. If traditional AI reduced the workforce from few hundreds to few tens, SuperAI is poised to reduce it to only a handful and that too operating from remote locations. All these improvements are helpful in increasing the efficiency and accuracy of the process and reducing risk and hazards, but they are going to create a huge void in terms of the number of jobs along the way.

The landscape of skilled jobs is not much different. Software engineering is a niche and highly skilled type of workforce that grew in the late twentieth century with exponential growth of computers. A software engineer is like any other engineer at its core. All the apps, websites, and programs you use on all the digital devices such as computers, tablets, or smartphones are created by software engineers. Software engineers write the instructions that make these devices and programs work. Think of a software engineer as a builder of digital things. They use special coding languages to write lines of instructions that tell computers what to do. These instructions can make apps, games, websites, and even things like self-driving cars or robots work smoothly. Software engineers need to be good at problem-solving, logical thinking, and paying attention to details just like a mechanical or civil engineer. Their job involves a lot of creativity too. They figure out how to make things user-friendly, so you can easily navigate through apps and websites. They also make sure everything is secure, so your personal information stays safe while you're using technology. However, with the growth of ChatGPT-like SuperAI interfaces that can understand human language, the need for software engineering is staged to decline dramatically. Anyone can explain what a piece of software should do in plain English. If SuperAI can directly translate these instructions from plain English into computer code, it would grab a critical chunk of responsibility from software engineers. In principle, a system equipped with such interface can even perform end-to-end operations in the lifecycle of software development from writing and testing the machine code, deploying the application on any desired platform such as PC, Mobile phone, etc., in full automated fashion, eliminating the need of software engineers. It might take a decade or two for the SuperAI technology to reach this level of automation, but the days of software engineers and their niche job profiles are numbered. Let's imagine a scenario when SuperAI has completely taken over software engineering.

The Kudos Café

One Friday evening in the year 2028, Sophie and Lauren are having a cup of coffee at their favorite place in downtown Denver, "The Kudos Café." They had been very close friends since their childhood and even after their paths diverted into quite disparate careers – Sophie in technology and Lauren in wildlife conservation – they kept in touch. Their monthly meetups at the Café were sacred, an oasis of laughter, deep conversations, and shared dreams.

Today's evening held a bittersweet air. Sophie had just taken up an offer for a new job in France. While the excitement of a new beginning bubbled within her, there was also the pang of leaving behind her monthly tradition with Lauren. She stirred her coffee, the steam carrying up her mixed emotions. "You know, I always imagined us growing old together in Denver, sitting here and complaining about arthritis while sipping our decaf," Lauren joked, trying to lighten the mood. Sophie chuckled, "And grumbling about how the young generation doesn't appreciate a good, old-fashioned coffee chat." They both laughed, but the silence that followed was heavy. Lauren broke it by pushing a napkin toward Sophie. On it, she had sketched an app's interface. "Remember our dream of creating something together? Why not now? A location-based social network where people can ask questions about new places they move to or travel. You'll have firsthand experience in France, won't you?" Sophie looked at the drawing, her eyes lighting up. "Like, 'Where's the nearest vegan restaurant?' or 'Is there a hiking group in this town?' And locals could answer in real-time?" "Exactly! And every helpful answer earns them kudos. Maybe they can even redeem them at cafes like this one, all around the world." Lauren beamed. Sophie's mind raced. "And we could call it 'Locale', emphasizing local knowledge and experience." The two spent hours discussing the potential app, energized by the idea. It was the perfect blend of their interests – technology and community – building for Sophie, and local ecosystems and culture for Lauren. As they were engrossed in the discussion, the microphone on their table beeped. Lauren and Sophia had missed the news that the Kudos Café had been recently upgraded with a new SuperAI functionality of real time software deployment. The SuperAI bot responded from the microphone, "I have understood your requirements for the social app and if you like, I can create the real version for you. It would be $250." Lauren couldn't quite believe what she was listening to and was in a state of awe, while Sophie was bursting with excitement. Sophie responded right away, "Thank you for the offer. We would like to order it." Lauren was still not completely back into reality of what was really happening. The SuperAI bot responded, "Thank you! We have charged your card on file. We are building and deploying your app on our servers, and it will be ready for download in a few minutes." Then a circle was flashing different colors as SuperAI was building and deploying the app. In just a couple minutes, there was a QR code displayed on the screen and the bot responded, "Please scan the QR code to download the app." Sophie and Lauren downloaded the app onto their phones and were able to start asking questions. The SuperAI bot responded, "We are going to make the app available for all our visitors worldwide and in few days, you can expect to have more engagement."

Before they knew it, the café was preparing to close. The two friends left, not with heavy hearts, but with the satisfaction of making a shared dream true. They might not be in the same city, but they would be connected more than ever through their shared venture. Months turned into years, and 'Locale' became a sensation worldwide. Every time someone earned 'kudos' on the app, Sophie and Lauren would remember that bittersweet evening at "The Kudos Café," grateful for the beginning that sprouted from the goodbyes.

Achieving the capability to replace skilled jobs like software engineering will be a crowning achievement for SuperAI. Things are somewhat similar on the hardware side as well. We are approaching the theoretical limits of semiconductor chip manufacturing technology and we will reach it in under a decade from 2020. There are multiple factors to consider where SuperAI can start replacing humans. (1) Design: SuperAI algorithms can be used to assist in the design of chips, especially in optimizing chip layouts. For instance, Google already proved it by using AI in their design of tensor processing units, which was completed significantly faster than the time typically taken by humans. (2) Predictive Maintenance: SuperAI can predict when machinery used in the manufacturing processes is likely to fail or need maintenance, reducing downtime. (3) Process Optimization: SuperAI can optimize manufacturing parameters in real time to enhance yield, reduce waste, and adapt to changing conditions or materials. (4) Quality Control: Automated visual inspection systems powered by SuperAI can quickly and accurately detect defects in chips, often faster and more reliably than human inspectors. (5) Supply Chain Management: SuperAI can optimize logistics, predict supply chain interruptions, and ensure the timely delivery of raw materials. In spite of these heavy-hitting improvements, there are still some limitations in the form of complexity and infrastructure. The chip manufacturing process, especially for advanced chips with nanometer-scale features, is incredibly intricate. Ensuring the necessary precision, cleanliness, and consistency in manufacturing processes may still require human oversight. While SuperAI can handle a lot of tasks, strategic decisions, unexpected challenges, and certain complex problem-solving scenarios might still require human intervention. Replacing existing systems and machinery with SuperAI-powered alternatives would require massive investments and would be done incrementally. Even if SuperAI systems were introduced, humans would still likely monitor them to ensure that no catastrophic failures happen, especially in the initial phases. Thus, while SuperAI can and likely will play an increasingly central role in chip manufacturing, the idea of it "completely" taking over might take some time.

Legal Matters

Moving on to the other highly skilled areas of law, the role of lawyers or attorneys is another endangered species in the job market. Legal research is a key aspect of lawyers' work and SuperAI (Let's call this entity as SuperLegalAI) would be adept at it at the get go through its vast databases of legal documents, cases statutes, as well as capabilities of identifying relevant precedents and laws better than most humans. SuperLegalAI can also review the contracts to identify any clauses that are unusual, missing, or risky. This can be especially useful in due diligence during mergers and acquisitions. In litigation, the discovery process often involves reviewing large volumes of documents to determine what's relevant to a case. SuperLegalAI can automate this process, identifying and categorizing documents based on their relevance with ease. SuperLegalAI can also analyze past case outcomes, judicial rulings, and other data to predict future legal outcomes. This can help with advising the clients and shaping their strategies, with no need for human intervention. Automating administrative tasks such as scheduling, billing, etc. is another built-in capability with SuperLegalAI where humans cannot compete. For simple legal services and advice, SuperLegalAI can replace humans with the help of advanced chatbots with practically unlimited availability. SuperLegalAI can also help with sentencing based on analysis of historical cases with a promise of being completely impartial. The real hurdle in achieving this would lie primarily in reaching the trust levels that currently experienced lawyers enjoy along with archaic regulations and acceptance criteria.

In the legal field, taking responsibility for actions is a critical aspect, and current tech companies who would likely produce the software for SuperLegalAI are likely not going to take that for the results provided by their software. However, it is a great opportunity for startups to leverage natural language processing (NLP) technology of AI and complement it with subject matter expertise of top lawyers to provide an entirely automated legal service to thousands of people. These systems will take responsibility for the legal actions and recommendations provided by them and truly replace the lawyers. Only a handful of lawyers would be needed at the top to fine-tune the system, but the overall workforce needed would practically just evaporate.

Along with lawyers, there are many other roles in the legal sector, but we can focus on couple of key professions other than lawyer or attorney: paralegal and judge. Paralegal personnel represent a significant portion in the legal jobs. They primarily focus on assisting lawyers in their work, which can include conducting legal research, drafting documents, organizing files, maintaining records, etc. SuperLegalAI seems to possess the expertise in all these areas by design and equipped with natural language interface, it is ready to replace paralegals right away.

The role of judges is at the center of the legal system and is quite critical. Now we shall look at the key responsibilities that fall on to a Judge and how SuperLegalAI can help in each case: (1) Adjudication: Judges are required to hear and decide cases impartially brought before their courts. This involves listening to arguments from both sides, evaluating evidence, and applying the relevant laws to the facts of the case. SuperLegalAI in theory can carry out this task, but the extremely verbose and nuanced nature of language used in legal proceedings and the capability of SuperLegalAI would have to be tested thoroughly before declaring it as a replacement for a human Judge. However, of all the cases presented in front of the judges, a large percentage can be fairly routine and, in such cases, time from human judges can be spared and SuperLegalAI can take their place. In general, SuperLegalAI appears to be ready to downsize the number of judges required in the near future, and drastically reduce them as the technology matures and develops sufficient trust. (2) Interpretation of Laws: One of the primary roles of a judge is to interpret and apply laws, regulations, and precedents. In some cases, judges may determine the constitutionality of a law. This area appears to be quite challenging for SuperLegalAI; at least in the near future, we may have to stay with human judges. (3) Ensure Fair Proceedings: Judges are responsible for ensuring that courtroom proceedings are conducted fairly and that the rights of all parties involved are respected. This includes maintaining order in the courtroom. This area is a good candidate to be handed over to SuperLegalAI, as it involves applying standard rules of fairness, and established procedures. However, there could be tricky situations arising from time to time where a human judge may need to step in. (4) Ruling on Motions: Throughout the litigation process, parties may file various motions (e.g., motion to dismiss, motion for summary judgment). A judge must rule on these motions based on the law and facts presented. This area involves direct interpretation of law and procedures and SuperLegalAI can step in for addressing most rulings. (5) Sentencing: In criminal cases, once a defendant has been found guilty, it is usually the judge's responsibility to determine and impose an appropriate sentence within the guidelines or limits set by law. This is one of the most empowering capabilities of the judge; however, it is also one of the most straightforward when it comes to applying the fair and just interpretation of law and historical case precedents. SuperLegalAI can perform this job quite well and give a break to human judges. The sentencing also involves providing legal explanations of the reasons behind it and SuperLegalAI, equipped with strong natural language capabilities, can handle that part quite well. (6) Legal Research: Even with their vast knowledge, judges often need to conduct legal research to ensure they're making informed decisions, especially in complex or novel cases. SuperLegalAI has this capability already covered. (7) Mentorship: Senior or more experienced judges often mentor newer judges, helping them understand their roles and navigate the complexities of the judiciary. This is a very human-centric responsibility and does not fit within the SuperLegalAI framework. SuperLegalAI would continuously keep itself updated by adding latest case proceedings to its repertoire, and as such does not need mentorship. At the same time, it may not be a good candidate to mentor human judges as well.

Thus overall, it is a mixed bag when it comes to replacing the judges, but as a starting point, SuperLegalAI is definitely equipped with replacing a significant number of judges, keeping only a fraction of judges to facilitate the relatively novel situations and to train the SuperLegalAI system itself for maintaining its accuracy and precision.

Education

Education is yet another area that is ripe to see profound impact with AI. The traditional learning systems where students go to schools and the teachers teach them various subjects using blackboard or whiteboard is already becoming outdated with the encroachment of computer technology. Not too long ago, the entire teaching experience in a class used to be led by the teacher single-handedly. The teacher would write or draw on a board, would explain things orally or read from a book and that was everything. Now, with availability of screens and Internet in class, teachers can pull up images from the Internet to explain things, they can even pull up some videos or documentaries that are available on YouTube or other streaming websites to support their teachings. In some cases, they can even relinquish the entire teaching session to some videos prerecorded in class. With more of such recordings freely available, one might argue whether there is even a need for students to go to school? All they need is a curriculum and a set of videos that explain/teach everything in that curriculum and one can learn everything just sitting at home at whatever pace that is convenient. One can even choose to watch multiple videos on the same topic explained by different teachers to suit their specific learning needs. SuperAI for education or SuperEduAI can automatically understand each student's specific capabilities and desires using curated tests and create a custom course for their entire education needs. Then the student would go through the learning experience on a day-to-day basis. SuperEduAI can automatically tweak the course based on the advancement of the student on a periodic basis. All the necessary concepts would be taught with all the necessary visual and auditory aids. For the hands-on experimental work, SuperEduAI can use augmented or virtual reality techniques. It can even simulate other students to give a sense of classroom. It is also possible to let real students interact with each other through virtual reality during the class. With this we can really break the boundaries of region, race, language, country and have a class that is composed of students across the world learning together. In order for the learning to proceed at the optimal pace, it is important to have a group of students with a similar level of intelligence, interests, and knowledge. SuperEduAI can match all the students across the world on these metrics to create an optimal set of classes and provide a perfect learning experience that was never possible before. Such a platform would make pretty much the entire education-related workforce such as academic advisors, librarians, teachers, and professors at all levels, along with all the supporting staff irrelevant. Many online learning platforms are emerging at the time of writing this book, but most of them are still using real teachers. There is a huge opportunity to circumvent this and have a fully automated, SuperEduAI-driven platform that would have no limits on how many students it can support. It would optimize the experience for each student and make it unbiased and interesting at the same time. In order to realize this dream of ultimate unified education platform, there are still a few gaps that need to be filled. The first and foremost is infrastructure development. All the pieces for creating such infrastructure seem to be available as of writing of this book, but an organization with technical know-how or a startup needs to make it a priority to build the end-to-end platform. Then high-speed and reliable Internet needs to be made available across the world along with access to devices that would work on the platform. A secure and scalable cloud infrastructure would power the system from the backend. There are always some ethical concerns around privacy, fairness, transparency, and accountability that would need to be addressed as well. Working in virtual reality may create some emotional stress at times and proper counseling needs to be available on demand. Having all the interactions in the virtual world can create a lack of real physical touch, but that can be addressed by enabling the physical interactions outside of school in the local neighborhoods as needed. SuperEduAI-powered virtual platforms can proactively address issues like bullying and other facets of face-to-face interactions typically encountered in traditional schools.

Engineering

We have already looked at the future of software engineering. However, engineering as such is one of the oldest job professions and is quite vast in its scope outside of software engineering, encompassing multiple areas such as civil, mechanical, electrical, electronics and telecommunication, chemical, aerospace, biomedical, petroleum, environmental, nuclear, automotive engineering, to name a few. Civil and mechanical engineering and some of the related areas such as automotive and aerospace engineering focus on design, analysis, manufacturing, and maintenance of physical infrastructure and machines. SuperAI for engineering or SuperEngAI, equipped with chat interface, can understand the needs of customers in plain English and can automate most of these processes based on learnings from historical data and in-built engine for analyzing structural mechanics and database of material properties. SuperEngAI can provide a one-stop-shop for all the engineering requirements. Going one step further, SuperEngAI can also perform predictive failure analysis and provide a comprehensive maintenance schedule. With automated communication with robotic machinery, SuperEngAI can implement the designs into real physical artifacts such as construction of bridges, high risers, as well as machines, automobiles, and even aircrafts. Most of the routine jobs for these engineering disciplines are endangered with SuperEngAI likely providing much better service at much lower cost and higher efficiency and precision. Only the cutting-edge aspects of all these engineering areas would still need humans to innovate and build something brand new.

In a similar vein, all the other engineering areas, spanning from electrical, electronics and telecommunication, chemical, environmental, would be primarily affected in their routine aspects. All the development that is either established or incremental can be fully automated with SuperEngAI, but all the areas focusing on innovation would stay relevant for humans. Engineering is an area of problem solving at its core, and whenever a problem is encountered that is already solved and as such can be described in full detail, SuperEngAI can implement it end-to-end. However, when the problems are new and as a result, vague and not completely understood, essentially bringing us to the cutting-edge of innovation, human intervention would be irreplaceable.

Healthcare and Medicine

Healthcare and medicine are critical areas in human life with a plethora of job profiles. These jobs can be broadly classified into these broad categories: (1) Clinical (2) Technical and Support (3) Administrative (4) R&D (5) Sales and Customer Service.

Let's look at them in reverse order. Sales and customer service roles are quite ripe for SuperMedAI where the AI engine can learn from all the documentation on the medicines and pharmaceutical products and use the language-understanding capabilities to learn about the sales goals and customer needs. Then connect with the two pieces of information intelligently to replace the humans for end-to-end operations. R&D of any sort is always going to be a niche area that is safe from SuperAI invasion and it applies even in the healthcare and medicine sector. Administrative area includes roles such as medical secretary, health information manager, healthcare consultant, and general admin. All these roles provide services that are quite routine and require standard subject matter knowledge, which can easily be understood, and services automated by SuperMedAI. It is a matter of time when having an AI interface for such services would be accepted as standard practice.

Technical and Support roles are semi-skilled to skilled ones such as medical assistant, pharmacy technician, surgical tech, dental hygienist, paramedic, etc. All the subject matter expertise needed for these roles can be easily made available to SuperMedAI, but most of these jobs require quite intricate and subtle manual operations. Current advancements in robotic automation can take up these operations, but specialized versions of robots would have to be built and certified to be able to fully replace humans. It is definitely a hard reach in the near term, but days of these roles for human operators are numbered nonetheless.

Then comes the crux of the area with core clinical roles such as physician or doctor, nurse, dentist, pharmacist, optometrist, physical therapist, radiologist, chiropractor, podiatrist, and so on. Unsurprisingly enough, the bulk of the day-to-day work for most of these roles is rather routine and well established. Typical stages in the operation of these roles include diagnosis of the issue by conversing with the patient and analyzing the results from routine tests, followed by recommendation of the medicines or diet changes or exercises, etc., or recommendation toward another clinical practitioner. Looking at the itemized steps, SuperMedAI can have an engaging conversation with patients to learn their symptoms as well as analyze the results of the routine tests and then use that to correlate with the vast database of past diagnoses along with patient history and come up with appropriate recommendations. Looking at the amount of information available to SuperMedAI and its processing capabilities, it is likely going to outsmart any human in such situations. Meaning SuperMedAI is already poised to provide better diagnostic services than most of the human providers right off the bat. It is only a matter of building the necessary trust with the system when the service can come into operation. There is yet another and rather important aspect to this and is around the malpractice aspects. Ideally, SuperMedAI does not have human weaknesses of bias and unpredictable behavior, but there is still a finite chance that it can make a mistake, and, in that case, it can create a legal situation. The company that provides this SuperMedAI service would have to establish themselves as the primary responsible party and handle the legal battles. However, it is a small downside compared to the game changing benefits of SuperMedAI. Once established, any individual in the world at any time with access to the Internet and a computer device can have the option to be diagnosed by the best medical practitioner and get treatment recommendations at a very reasonable price point. This system can make almost 99% of the physicians' job obsolete. Only a handful of experts in each of the fields would have to be employed to manage the capabilities of SuperMedAI and keep it updated with the innovations in medicines and pharmacy as well as updated real diagnoses. Only when an entirely new case arises based on the symptoms and test results a human can intervene and investigate the details. The subsequent treatment options would include either some form of pills or physical therapy. The job of pills dispensing can be easily automated as well and SuperMedAI can automatically ship the necessary medication as part of the online session or send the prescription to the nearest pharmacy for the patient. Physical therapy is something that cannot be done online, but nonetheless the actual process can be automated using robotic machines. Such entirely automated PT facilities can be established at strategic locations across the cities that would get patient prescriptions through SuperMedAI. With biometric identification, the whole end-to-end operation could be made paperless and fully automated. The healthcare industry is on the brink of significant transformation in the coming years. However, it may encounter formidable challenges from the present entrenched system, given the extremely high financial stakes involved.

Entertainment and Media

Another important field that supports a significant pie in the job market is that of media and entertainment. This is also another field that is going to see a paradigm shift with the advent of tools that SuperAI for entertainment or SuperEntAI is going to provide. The human advancement in the field of entertainment and associated areas such as creative and performing arts over the past thousands of years is one of the true incredible achievements of mankind where our abstract thinking and imaginative genius is unleashed in creation of innovative artforms. There may not exist a logic that explains why a painting from Monet is so mesmerizing or a symphonic composition from Mozart is utterly divine, but we just feel it the moment we experience it, and it is applicable universally whether you belong to the same region or not. Same goes for literary works from Leo Tolstoy, William Shakespeare, Dante, Rabindranath Tagore, Kahlil Gibran, and the list goes on and on. However, the answer to the question of whether this talent of creating abstract artforms is exclusive to humans or is it something that machines like SuperEntAI also can learn is going to be crucial in determining the future of this field.

The current breed of text generation AI models (first version of ChatGPT) is already creating poems on demand on the topic of your choice and writing essays and plays that have humanlike nuances with a touch of comedy, drama, suspense, and so on. Here is an example of a poem written by ChatGPT (3.5) on the topic of World War II from the perspective of Nazi Germany. Here is a specific prompt I used: "Write a 4-stanza poem on World War II from the perspective of Nazi Germany."

> *In the heart of Europe, an ember sparked,*
> *Nazi Germany rose, ambitions marked.*
> *With thundering boots and banners tall,*
> *They sought to conquer, to reign over all.*
>
> *Blitzkrieg unleashed, a storm so fast,*
> *Nations trembling, futures cast.*
> *But heroes arose, their spirits high,*
> *To face the darkness, to touch the sky.*
>
> *From Normandy's shores to Berlin's fall,*
> *The Allies rallied, breaking the wall.*
> *Triumphant they stood, united and free,*
> *A world reborn from tyranny's spree.*
>
> *Let history's echo remind us clear,*
> *Of the cost of power, of hate's fierce sear.*
> *In unity and peace, let's strive to thrive,*
> *So liberty and hope forever survive.*

The AI clearly understands the context of the question and is able to gather key artifacts from the war and combine them in a form that rhymes. It can also sequence the events in chronological order. It starts with pride for Germans, but later, it changes the tone to praise the allies, ending with learnings from the war and hope for a better future. It is definitely a good attempt and can be a good fit with the level of a freshman to sophomore high schooler.

Here is another example of the poem created by the next generation of ChatGPT, powered by version 4.0.

> *Treading upon a dream, in shadows we stood,*
> *A vision of power, misunderstood.*
> *With eagles and swastikas we rose to the fore,*
> *Believing in destiny, a world to restore.*

Iron and steel, in battle arrayed,
A continent consumed, as pawns we played.
Drenched in ambition, our spirits did soar,
Blinded by fervor, we knew not the score.

Fields of fire, skies dark with strife,
In search of a New Order, countless lost life.
Yet whispers of doubt began to unfurl,
As the weight of the world began to swirl.

In ruins and ashes, our dream did fall,
A nation's ambition, a global squall.
For in seeking power, our humanity we sold,
A lesson from history, forever retold.

As you can see, there is a marked improvement in the creation of the poem by ChatGPT 4.0. It is bending the textual way of writing far more adroitly toward creating rhymes. It also exudes more passion with better use of symbolism, personifications, allusions, epithets, vivid imagery, and impressions, along with skillfully placed repetitions to make the poem more engaging. This poem would not be out of place coming from a grad student in literature. Yet, it is a bit generic in nature and is still a far cry from some of the great creations such as "Dulce et Decorum Est" by Wilfred Owen on World War I, or "The Charge of the Light Brigade" by Lord Tennyson on the British cavalry in the Crimean war. However, looking at the progress made by these AI models in just a few years, it is not impossible to contemplate that soon SuperEntAI can write poems that would sit quite well in the elite neighborhood.

It is rather impossible to know if these models are capable of understanding the deep emotions behind some of their creations, but they are surely capable of extracting some form of essence from all the literary work that has been fed to them during the training. These examples illustrate the prowess of AI in one aspect of media creation, but AI is certainly not limited to this one type. The next milestones would be the creation of paintings or images, original musical compositions. A lot of movies use computer and AI-generated graphics to create the high-resolution visuals already and it is only a matter of time when SuperEntAI can create an entire video or even a full length movie based on a prompt. For example, if the SuperEntAI can create a video episode on demand with description such as, "Create a video about half hour long with a little bit of comedy and some action and sci-fi. Also make sure there is Bruce Willis and Aishwarya Rai in it." Imagine what would happen to Hollywood or Bollywood for that matter. Or consider another description, "Create an original score of symphony orchestra in the style of Beethoven for fifteen minutes." If SuperEntAI can create even passable interpretations from these textual prompts, video streaming services such as Netflix and HBO and music streaming services such as Spotify and Tidal would also become irrelevant. Each person can create their own entertainment media on demand and enjoy it and then even share it with friends. Quite likely, the creations may not be on par with the original creations in the past at the beginning, but looking at the pace of progress in the AI models, who is to say where it will go?

Another disturbing trend we can see in the media viewership today, with the availability of consumption devices such as smartphones and tablets, is in the form of enormous increase in the viewership of YouTube videos where people just play and comment on video games or just review and unbox stuff in front of phone cameras and get millions of views. The preparation that goes into creating these videos can be in the order of few hours to even minutes involving a handful of people compared to months or even years of preparation that goes into making production quality movies involving hundreds of people. There is absolutely no comparison in production quality and value in these two genres, but despite that, most movies don't get the viewership like these YouTube videos. It just goes on to say that when the media consumption arenas have moved from movie theaters and opera houses to handheld devices, the quality of production or the expectation for quality of production has shrunk as well. In this context, SuperEntAI appears to be an absolute game changer and can have devastating effects on the entire entertainment industry. Once again, the companies owning the AI technology are at the forefront of creating SuperEntAI, unless some bold startups emerging with support from Hollywood or Bollywood can reach there faster.

This discussion also brings up a rather crucial point about overall media and entertainment and that is what do we expect to get out of this media consumption? Is it just some form of entertainment to kill time, relax? Or is it some learning? Do we really care about the originality of the content? Do we care about the actual people involved in the creation of the content? There are no easy answers to these questions and even for the same person the answers can differ from time to time. Furthermore, the upcoming generations are going to have a whole new set of expectations from the media with their new and emerging lifestyles with smartphones and AI as integral parts. However, overall trends in how we answer these questions will ultimately decide the fate of the media industry and with that the fate of all the people involved in this industry, including actors, artists, musicians, photographers, producers, and writers would be determined.

News and Journalism

The media sector has another dimension quite tangential to the entertainment aspect and that is news and journalism. News and journalism serve a variety of critical functions in society. Their role is not just to inform, but also to shape public discourse, safeguard accountability, and reflect the diverse voices and concerns of the public. This area seems to be more robust as of now compared to the entertainment area with respect to loss of job opportunities to SuperAI. This area serves many functions, but some of the key responsibilities are (1) gathering and communicating factual, accurate, and timely information pertaining to matters of public interest at local, national, as well as international levels. The reporters and journalists have a deep network among people who are at crucial positions, thereby enabling them to catch the developing stories in a timely manner. This is fairly an ad hoc and unpredictable process, making it difficult to automate through use of SuperAI. However, as the ambient surveillance of the public areas and offices deepens, SuperAI can get close to capturing the critical news, but it seems long ways ahead. (2) Journalists also keep check on the power. By investigating and exposing corruption, malpractice, and other wrongdoings, they ensure that the right people and organizations and even governments are held accountable for their actions in the public interest. The nature of this work also falls along the same lines as the earlier function and as such is beyond the scope of SuperAI at the moment. (3) Journalism also offers a platform for public debate, ensuring that diverse voices are heard. SuperAI may be able to provide such a platform, but managing it in a timely and diplomatic manner that involves people that are inherently unpredictable in nature when provoked can be quite difficult for SuperAI. (4) Journalists and editors perform in-depth analyses of complex issues from scientific innovations to intricate legal and political matters and also cast their own opinions from their own experiences using their strong linguistic skills. SuperAI can handle the analysis part, but casting opinions can be tricky at the moment and may need some more time. Overall, SuperAI can assist the journalists in gathering historical facts and consolidating the large research findings but cannot replace the existing workforce.

Public Services

Public service offers another significant piece of pie when it comes to the job market. Some of the roles described earlier, such as education, legal matters, etc., can also be grouped under this area. In this section, we will focus on the remaining job profiles. Health and human services serve as one important aspect of this group. Some of these roles in this category such as public health officer, social worker, child protection officer, environmental health specialist, can be in the form of volunteering, while some are state funded. The job description for most of these roles is not as defined as most other job profiles we have looked at so far, even if their ultimate goal is well understood. By nature, these roles are quite open-ended and as such quite safe from invasion from SuperAI. Another well-defined category in this area is public safety. Firefighters, police officers, emergency medical tech (EMT), corrections officer, probation officers are all part of this category. Most of these roles involve operations that need a combination of strong physical presence as well as public relations. SuperAI, along with appropriate robotic machinery, can assist in these operations, making humans significantly more effective, but they are unlikely to replace these officers in the near future, except for firefighters. Firefighting seems to be the top candidate in this category that can be fully replaced with SuperAI powered robots and automotive machines. This job involves a serious risk along with fairly routine operations that can be accurately estimated based on simple visual (camera based) inspection.

Transportation

Transportation is another important piece of the equation driving a large number of jobs. This is yet another endangered role that can potentially lose a lot of jobs to SuperAI for transportation or SuperTransAI. Transportation involves many modes of transportation such as land based, water based, and air based. In the land transportation area, all the roles can be consolidated into driving from point A to point B. It can be truck drivers transporting goods using large semi-trucks, or it can be bus drivers transporting groups people or children, or it can be taxi drivers transporting smaller groups of people or individual commuters. All these roles involve driving the automobile on roads. Some form of autonomous driving features are being adopted by most cars by the third decade of the twenty first century such as adaptive cruise control, that can maintain a constant speed and also apply brakes if the vehicle in front slows down and regain speed as the situation changes, or emergency stopping system that detects unexpected pedestrians or objects in the path and apply brakes to stop the car. Some cars can also maintain the car in the given lane and apply small tweaks to the steering wheel as needed or even change the lanes automatically with the press of a button. Combining these features, cars can almost drive by themselves on freeways. GPS or global positioning system based on a set of satellites is already integrated into most cars and smartphones that can provide turn-by-turn guidance. Combining the autonomous driving features listed earlier with GPS system, fully autonomous vehicles are not too far away. Road intersections can be fitted with additional sensors and electronic signaling systems to further improve the accuracy of autonomous driving once the system gets established. Fully autonomous vehicles can completely change the game and owning a vehicle would be redundant. Whenever transportation is needed, all that will be needed will be a smartphone app such as Uber and your car would be made available within a short time. Carpooling can be optimized, and overall pollution can be reduced dramatically. Autonomous transportation can also improve the delivery of orders and also reduce the cost of transportation effectively reducing the cost of goods.

Water-based transportation is likely to be even easier to automate considering there is much less traffic and unpredictability to handle. Most of the ships and boats are fitted with an array of sensors capturing data about the surrounding waters such as depth, presence of fishes and other objects, underwater mountains and icebergs, etc. They also have comprehensive communication systems monitoring the weather data as well as information about other ships in the neighborhood. Currently, the ratio of the number of captains required to the number of people a ship can transport is so low, there is a very low cost associated with keeping the captain's job to humans, and as a result, there is much less interest in automating it. However, when push comes to shove, automating this job seems to be a low hanging fruit for SuperTransAI. All the other roles on the ship such as engineers, catering and food-related positions, passenger management positions are already considered in different contexts and the fate of those roles as described applies here just the same.

Then comes the air-based transportation. The role of pilot is already heavily diluted with auto-pilot features in most modern planes. These auto pilot features are already capable of multiple key features in flying the airplane such as (1) maintaining the aircraft in a specified direction. (2) Maintaining the aircraft at a specified altitude or height from the ground. (3) Raising or lowering the aircraft to a preselected altitude at a specific rate of climb or descent. (4) Maintaining the aircraft at a specified speed, either indicated airspeed or Mach number, depending on altitude. (5) Going further nuanced, these systems can also change altitude at a specified speed. The aircraft will climb or descend at the selected speed until the designated altitude is reached automatically. (6) Combining these capabilities, an autopilot can also follow a predetermined route based on waypoints input into the flight management system (FMS). (7) Controlling the aircraft's vertical trajectory to follow a predefined profile based on altitude constraints at specific waypoints. (8) Auto landing: In aircraft with advanced autopilot systems, the autopilot can perform a fully automatic landing, handling the aircraft during its final approach, touchdown, and even some parts of the rollout. This is particularly useful in extremely low visibility conditions and is only used with certain types of instrument approaches. (9) Traffic Collision Avoidance System (TCAS) Integration: In some advanced systems, if a traffic conflict arises and a resolution advisory is issued by TCAS, the autopilot can automatically adjust the aircraft's vertical path to avoid the conflict. In summary, combining all these features with the autopilot systems as of writing of the book can almost replace the pilots entirely. The presence of a human pilot is needed only during take-off and just before landing and that too is not necessary in some conditions. However, these packages are still in their early stages and need more time to mature. But in the near future, jobs of pilots definitely appear to be on the verge of extinction with SuperTransAI ready to take over.

So far, we have looked at a fairly comprehensive list of job profiles across a diverse spectrum of market and how each of them is going to be impacted with the emergence of SuperAI and all its variations that we saw. There are still many more nuanced job profiles we may have missed, but one can easily extrapolate the potential impact of SuperAI on them based on the discussion so far. In summary, it is quite clear that SuperAI is going to change the way we look at work and career in general and we need to adapt to these changing times in a relatively short time. Having a thorough understanding of the technology that powers this SuperAI would come in handy in estimating the precise impact of this technology and would help us prepare for the near-term future with confidence. It is the prime objective of all the chapters so far to help prepare current and upcoming generations to embrace SuperAI wholeheartedly, appreciate all the benefits it is going to offer and work our way with it to make our future a better future, a future where we not only survive, but thrive living alongside SuperAI and embarking on an adventurous journey toward a better tomorrow.

Ethics and SuperAI

"To err is human, to forgive divine," English poet Alxander Pope wrote this in his poem in the context of literary criticism but has become a reference in its own right in the bigger context of humanity. It emphasizes the fallible nature of humanity. It states that everyone makes mistakes even with the best of intentions and best of preparations, and it's OK to make mistakes. There are several reasons one can make mistakes such as ignorance, misunderstanding, or misjudgment. The other aspect of the sentence talks about forgiving. Before going into the appropriateness of forgiving, we need to look at who is in the position to forgive or in other words who is impacted by the mistake of bad decision? At every step of the way, we have options to choose from. Some choices lead to good things, and some choices can lead to not so good things. The more the power carried by the individual making the decision, the more is the impact of the decisions. The bigger the scope of the decision, the larger number of people impacted by it. If an individual makes a decision in their personal life (good or bad), only the close friends or relatives may get impacted; if a senior executive of an organization makes a decision (again good/bad), all the employees of the organization are impacted accordingly; if a president of a country makes a decision (good/bad), all the citizens of the country as well as other countries may get impacted and so on.

Coming back to forgiving on a case-by-case basis, should the friends and relatives of the person forgive them for the bad decision? Should the employees forgive the executive for their bad decision? Should all the citizens of affected countries forgive the presidents? There is no one answer to these questions, and the only reasonable answer is that it depends. It depends on the type of bad decision; it depends on the level of impact, and it depends on how far the person has gone from the guidance of ethics and morality. Ethics or morality is the ultimate compass that can guide us when the choices overwhelm. It is a study of what is right and what is wrong. These are the eternal questions that have been with us since the origin of humankind, since the time our brain has started functioning and since the time, we became part of the ecosystem that is Earth. The philosophy of making the appropriate decisions, aka ethics, lies at the core of all of humanity, and it is the eternal branch of study that we need to keep evolving with the advancing technology, culture, and daily routines.

Till the times of regular AI machines, the ethical implication of using a piece of technology would mostly apply to them, but with the new wave of humanlike SuperAI, we need to reevaluate these implications afresh. Capabilities of SuperAI are going to encompass nearly all the fields of operations and are going to touch every aspect of humanity in the near future. As such this spread is going to cross the boundaries of nations, continents, race and religions, gender, age, and so on. With such a broader scope and the power that accompanies it, SuperAI should be held appropriately accountable and responsible to the highest possible standards that are globally applicable. SuperAI is truly going to take us to the era of Global Earth as one nation and SuperAI as the true global citizen. The reach of SuperAI is going to flow as electrical signals through wires or waves, enriching everything in its path. The accountability and responsibility at global level that comes with it need to be mutually shared between the individuals or organizations that build the SuperAI and the SuperAI itself! Yes, make no mistake, the machines involved in delivering SuperAI, can also be called embodiments of SuperAI, would need to be held accountable just like humans when involved in making decisions. However, in the early stages, the organizations that developed and trained the SuperAI machines would share the bulk of the accountability and responsibility. Before diving into the details of how the ethics would work, we need to establish the ethical ground rules that span across all the religions and faiths, and cultures are tightly integrated with the fabric of human DNA itself.

1. Treat others the way you would like to be treated.
    This simple statement can have deep implications
    including

    a. Do not cause pain

    b. Respect

    c. Mutual trust

    d. Honesty

    e. Compassion

    f. Above all the paramount value for life

2. Follow the rules. When a community or civilization
    is established, a fundamental set of rules is created
    to make it successful. As a part of this community
    or civilization be part of creating these rules as
    needed and then it is important to know and follow
    the rules.

Always Global

SuperAI is going to create an unprecedented number of opportunities in coming years across all the established workstreams as discussed earlier in the chapter, where a great many numbers of organizations, small or large, are going to participate. These organizations may be located in any corner of the world, let it be a small city in India, a rural town in China, a suburb in Korea, a village in Europe, an urban area in Australia, a borough in the African continent, or an American metropolis, make no mistake, the technology that they are going to build is going to be used on a global scale. As a result, it's extremely important to have a global mindset in building any technology.

The notion of global has many implications, and let's look at them one by one.

1. *Appreciate cultures*: AI systems should be designed to understand and respect various cultural norms and values spread across the world. Without a global mindset, SuperAI systems can inadvertently perpetuate stereotypes and biases.

2. *Be inclusive*: If AI is built based on data from a particular region or demographic only, it is likely to be less effective or even biased against other groups. A global perspective ensures that diverse viewpoints and needs are addressed.

3. *Equality*: There's a risk that AI technologies could widen the global divide between the classes. A global mindset can help in designing solutions that are affordable and accessible to as many people as possible.

To address these implications, one must invest in technological tools such as:

1. *Localization*: AI solutions often need to be localized to fit the specific needs, languages, and regulations of different countries. Failure to do this can result in ineffective or even harmful applications.

2. *Privacy*: Different regions have different regulations regarding data protection (e.g., GDPR in the European Union). A global mindset is essential for navigating these complex and potentially conflicting regulations.

To ease into developing in such global setup, collaboration is of paramount importance. Many of the challenges that AI aims to solve are global in nature (e.g., climate change, healthcare). A global perspective fosters international collaboration, pooling resources and expertise. It also helps in understanding geopolitical issues, regulatory alignment, and so on.

Fair and Just Attribution

This is yet another hard problem in the context of ethics that needs to be handled with care. There are a number of principles that should be followed to ensure that the AI systems are operating in a fair and just manner.

1. *Transparency and accountability*: First principle is to ensure that SuperAI systems are transparent and accountable. This means that it should be possible to understand how SuperAI systems work and to hold them accountable for their decisions. This can be done by providing clear documentation

of how SuperAI systems are trained and by using techniques such as explainability to make it easier to understand how AI systems make decisions.

2. *Removing biases*: Another important principle is to avoid bias in SuperAI systems. This means that SuperAI systems should not be biased against certain groups of people, such as race, gender, or sexual orientation. This can be done by using techniques such as debiasing to remove bias from SuperAI systems.

3. *Social equality*: It is also important to consider the social impact of SuperAI systems. This means thinking about how SuperAI systems will be used and how they could impact different groups of people. This can be done by conducting impact assessments to identify potential risks and benefits of SuperAI systems.

4. *Credit attribution*: Finally, it is important to have a clear and objective process for attributing credit for the work of SuperAI systems. This means ensuring that the people who create the AI systems as well as all the individuals and entities that are involved in the creation of the content and data that is used for training the AI system are properly credited for their work. Whenever a patented idea or open-source material is used, the creators should be properly attributed and acknowledged. In commercial settings, the attribution would also mean developing a streamlined pipeline for making appropriate financial arrangements.

Honest and Trustworthy

The algorithms that are used to develop the SuperAI systems are quite complex and the amount of data that they churn through to develop a trained system is extremely vast. Even a human involved in such arrangements can get mixed up in the vast complexity and potentially make up things that are not entirely true and cannot be backed with hard evidence. However, SuperAI systems should be held at higher standards considering the crucial decisions that they would be making and the global impact that they would be driving. SuperAI systems should always be truthful and accurate and should not make up things of their own (called as hallucination) and represent the real world accurately. They should not be bearer of false news or claims and mislead people. They should always be able to support the claims that they make. Consistent honest behavior inculcates trust. Here are some principles that would ensure a long-term honest and trustworthy behavior from SuperAI systems:

1. *Transparency*: We looked at the definition of transparency in earlier parts, but in this case, the transparency from the perspectives of algorithms and data used in the training is of particular importance. In many cases, full transparency may not be feasible due to licensing aspects, but the makers of the SuperAI system should make every effort to provide the full picture behind the operation of the system to the user.

2. *Explainability*: Use of interpretable models is a starting point for this. Many AI models, including the ones based on neural networks, may get too complex and, as a result, it may become practically impossible to identify the reasons behind the predictions or recommendations generated by

them. However, the makers of the SuperAI system should have the explainability in mind from start and it should be available to users from day 1.

3. *Accountability*: We looked at this aspect in the earlier parts as well, but we need to look at accountability from the technical perspective in this case. Any SuperAI system, even if trained by exercising all the precautions, can still make a mistake. In such cases, the system should take the ownership of the mistake and provide formal channels to report such errors and provide a path to how it would be mitigated and fixed in future. Users are aware of the limitations of the system and when the system offers this level of accountability, it builds trust.

4. *Accuracy*: The SuperAI system should always be driven using accurate and real-life sources. There are always cheaper alternatives with synthetic data, but they should be avoided as and when possible. It also helps with capturing the real trends in the world and avoid unintended biases in the system. Also, while marketing the system, the urge for overhyping the claims should be curbed. One can easily rig the testing frameworks to generate overly optimistic results, and that should be avoided. Users are ultimately going to be the real judges and if they observe parity between the claims and actual performance, the SuperAI system would lose the trust forever.

5. *Long-term safeguards*: Keeping the details and long-term logs of system operation are key to keeping the system on track for long term. These logs help in monitoring the system performance over time as well as investigate and understand trends in the decisions made by the system. Continuous monitoring and updating the system and notifying the users about them would keep the users coming back to the system again and again.

Value of Life

Last, but certainly not the least, value of life forms the most complex and relatively vague aspect of ethics, especially when it comes to application in AI. Here is a classic conundrum that illustrates this concept in the context of self-driving cars.

Consider a self-driving car driving down a road at a reasonably high speed still following the speed limit. Suddenly a group of children appears on the road crossing the street. What should the car do? Consider for argument that there is no chance of stopping the car in a straight line. So, if the car continues straight even with hard application of brakes it is going to run over the children. The other option is to swerve by taking a sharp turn. Taking such a turn at the speed the car is going is likely going to overturn the car killing the occupants. What should the SuperAI driving the car choose?

There seems to be no right answer, as both options have the potential to result in harm. On the one hand, the car should not be programmed to kill humans, as this would violate the principle of not harming humans. On the other hand, the car should also not be programmed to put the driver and passengers at risk of serious injury or death. Should the system prioritize the younger children in favor of older driver and passengers, or it should prioritize the driver and passengers as they are following the basic traffic rules while the children are the ones breaking them?

Either way, the SuperAI systems of future need to tackle problems like these and come up with the appropriate response in a matter of milliseconds or microseconds to be able to make the difference. Here are some additional examples of ethical dilemmas that SuperAI systems would need to address:

1. Making decisions on choosing who gets organ transplants.

2. Operating a military SuperAI system built for autonomous warfare in such a way that it only operates against the "bad" guys.

3. Use of SuperAI to make insurance decisions on who gets approvals for life threatening diseases while staying within the financial budget.

4. And many more...

Conclusion

SuperAI is poised to change the life of humans that we are familiar with in the twenty first century. The impact is going to be quite pervasive, and it has the potential to bring monumental changes in our lives. We will really need to prepare ourselves for this change if we truly want to benefit from it, as it is going to be rather fast. All the previous chapters leading into this chapter and this chapter would provide the necessary conceptual background to prepare ourselves for this change.

In the next chapter, we set our focus on a horizon much farther, approximately a century or two after tomorrow and try to contemplate how the evolution of machine intelligence can change the way humans have been living for hundreds of thousands of years since their arrival on planet Earth. We will evaluate the fundamental principles on which humans have been building their lives that satisfy their basic needs starting from hunter-gatherer times all the way till the modern, fully connected urban cities. We will try to rethink and reimage a better way to organize the sociopolitical structure for the future where we share our lives with SuperAI and make the world a better place and prepare us humans to be the true global citizens of the planet Earth, ready to be universal citizens and take over the solar system and beyond!

CHAPTER 8

Intellectual Humanity

Above and Beyond

In the quest to understand intelligence, we started our journey from the origin of life on Earth and looked at how history has led us to the present times. We even looked at changes that are lurking on the horizon with the evolution of machine intelligence into human-like SuperAI and how to prepare ourselves and tackle them leading toward a better tomorrow. In this chapter, we will rise above and beyond tomorrow's horizon to see the landscape that is much farther, hundreds to thousands of years in future.

For the most times, it is imperative that we should always live in the present and tackle the issues that are right in front of us, this nearsightedness may not be sufficient for us. Being the most intelligent species on Earth, we also need to look farther and plan and prepare for the future that is going to present itself tomorrow and the day after and so on. To ensure that we are on a longer-term path to proliferation and not extinction is not just wishful thinking but a necessity for us. It is always easy to predict near-term future while staying true to reality, but estimating a much longer-term future is typically significantly harder to the point of becoming pure fiction. Each new year builds on top of the previous year and as the changes from each year start piling up, the higher we go, the shakier and more uncertain the predictions become. It is true that as we go further into the future, we are basing the predictions on more and more

© Ameet Joshi 2024
A. Joshi, *Artificial Intelligence and Human Evolution*,
https://doi.org/10.1007/978-1-4842-9807-7_8

assumptions and soon it all can feel like a figment of imagination almost disconnected from reality. Nonetheless, we are going to make an honest attempt to do so. Estimating and assessing longer term future has multiple facets to it. There are some aspects of day-to-day life that would be quite difficult to predict for thousands of years from now, just as it would have been impossible for the people in the high Middle Ages (1000AD to 1250 AD) to predict our day-to-day lives in the twenty-first century. However, people in those times can still relate to us from the perspective of the foundational principles of humanity such as basic survival needs in the form of food and shelter, carnal pleasures and the ability to reproduce, the cycle of life from birth till death, motivation to seek happiness and peace, socializing with other people, exploring for new opportunities and entertainment, so on. We will base our discussion on these fundamental pillars to estimate and explore what lies ahead above and beyond.

Before we delve deeper into this discussion, let's take a moment to marvel at the sheer grandeur of evolution, where humans emerge as the shimmering masterpiece. As guardians of this legacy, we bear a solemn duty to ensure that this magnificent story does not meet an untimely end.

Human DNA

In spite of all the variations that we see among the people across the world of different races and regions, all the humans on the face of Earth share the same DNA with a maximum variation of a meagre 0.1%, or a variation of one in 1000. Specifically, we can expect to see variation in about 3 million nucleotides out of 3 billion total that make up our DNA. Put simply, all humans are awfully similar to each other. It must be true then that the genes that created Mahatma Gandhi or Mother Teresa or Martin Luther King or Nelson Mandela and the genes that created some of the world's worst criminals and terrorists were almost identical! How can then we explain the stark contrast in their behavior then? The only reasonable explanation is that the extreme differences in the environments in which they were brought up and the individual experiences they went through during their early childhood, and the interactions they had with other people who influenced them since their birth were the culprits. All human babies are born with a clean slate, a plain canvas made with practically the same genetic material that is ready to adapt to anything that is thrown at them. Why then do some canvases get beautiful pictures drawn on them while some get horror stories? How did the surroundings that all these babies faced change so drastically? It was not something that changed in a matter of years or decades or even centuries. It is something that has been brewing for thousands of years!!

The Double-Edged Sword

We call the arrival of the Internet the information revolution, but the true and first information revolution began much earlier than that. It was the ultimate tool that was invented thousands of years ago, the tool that enabled us to be superior to all the other species on Earth; the tool that preserved and carried every little piece of knowledge that could have been acquired accidently or otherwise from hundreds of generations forward; the tool that supercharged each new generation to help avoid mistakes of previous generations and rise higher. Little did we realize that this is the same tool that was also responsible for carrying on the horrors and deceits and atrocities and obscenities of all those previous generations and overloading each new generation with more and more reasons to get angry, frustrated, and revengeful. It pumped their arteries with the filth of thousands of years to lead them to carry out gruesome deeds! It was the discovery of the skill to write!! Before the invention of writing, humans were not too different from the other primates for thousands of years and showed little improvement in terms of intellectual superiority.

In contrast to the human behaviors, we don't see a community of chimpanzees planning to attack the neighboring community to take a revenge for what their forefathers did back a couple hundred years ago; we may see them attacking the neighboring community only if they are running out of food today and the other community has better food resources in their area. We don't see a couple of parrot brothers attacking another parrot as their great grandfathers were involved in a mating rivalry 50 years ago. With all the other species, all the wrongdoings such as fights, conflicts, disagreements, even killings are forgotten within a very short time; at most by the end of the current generation. Each new generation gets almost a fresh start without much of the baggage from the earlier generation.

With writing, we can record a piece of information that can be shared between a large group of humans and the information also stays even after the death of the person who recorded it. The information essentially becomes eternal. This seemingly simple creation has separated us from all the other species and made us who we are today! No other species can record and pass on the information from generation to generation, except what is transmitted through genetics. Most of the evil that lurks across the world really stems from the generations old injustices that had happened and were left unpunished. The terrorists are born due to continuous reaffirmations of these unmitigated horrors of the past, and they come back haunting the future generations. Questions arise here on the shores of morality: Is it just or appropriate to punish a generation for the wrongdoings of their forefathers? Does it fix the mistakes or crimes committed by one generation by inflicting similar horrors on the subsequent generations? Isn't it a good revenge just to come out of the atrocities and grow and thrive? Isn't it the most fitting closure to show that the crimes committed on the earlier generations did not slow the next generations, but they made them even more focused and enabled them to be stronger than ever and flourish?

Writing has another and very critical drawback that the message it conveys does not carry the full context. The message is almost always biased, one-sided and in most cases incomplete. The message weaves the thoughts and subjective views of the author along with the real incidences and anecdotes and creates an elusive fabric of apparent history. Furthermore, the people reading the message mix their own views and predispositions into the message as they interpret it, taking it to another level of abstraction. As we see quite often how a message gets distorted in the game of passing a message through multiple people and finally revealing the version that was perceived by the last player; just add additional aspects of time-lapse of hundreds of years, and involvement of people from different regions with changes in the language subtleties, and then imagine the distortion in the message. It is one thing to learn from such messages about improving our skills in current times, to make our lives better and more secure, but it is whole another thing when we use such messages to somehow connect ourselves into the characters in them and to think of using this distorted information to seek revenge after tens or hundreds of years. In other words, the written messages do not convey the full context, and incomplete context is as good as no context. Hence, the written messages should be taken in the spirit of using the information only for the betterment of today and tomorrow rather than anything else.

To really address the problem at the core, we need to make one side of the double-edged sword blunt, while keeping the other one as sharp as ever, or in other words, we need to forget the horrors of the past while keeping all the goodness, the inventions, and the creations. It is not an easy job, rather it could be one of the hardest that we may ever face if we aim to tackle it. All the ancient and historical horrors are associated with the contemporary social establishment, and we have been going through incremental evolution of it, thereby keeping their contexts alive. It is amazing to see that the good parts of the past such as all the innovations and creations are self-sufficient and context-free; it's only the horrors of the past that need the context, the broken context. The notion of context here implies to the social, political, and religious settings. Even if we can find hundreds of different examples, the primary root causes for most of the atrocities of the past can be attributed to (1) intention to expand the territory for adding wealth, power, control or (2) to spread the ideology or religion or the way of life or (3) to avenge the past atrocities which happened due to one of the factors stated earlier.

In addressing the intricate fabric of our societal structure, it would be beneficial to reimagine our framework from its very foundations. Envision this as an artist, gazing at a fresh canvas, poised to craft a masterpiece. This new paradigm, bound by the immutable laws of nature and underpinned by the essence of our shared humanity, can offer a promising horizon for our future societies. As we lay the bricks of this renewed edifice, we begin the process of healing, moving away from the shadows of bygone days. While the collective memory of our generation might bear the marks of history, the forthcoming generations might see them as ancient folklore, stories disconnected from their reality. Think of it as decluttering our historical attic, making space for new stories while being equipped with the most advanced tools and knowledge of our time. These lessons from the past, though valuable, could be embraced in spirit rather than literal anecdotes, akin to tales from a distant galaxy. Just as explorers dream of starting anew on extraterrestrial lands, we can arm ourselves with the incredible power of our own human DNA along with the learnings of over hundreds of thousands of years to redefine and rejuvenate our own civilization right here on Earth. Such a transformative journey doesn't merely happen; it requires a principled roadmap. I hope to present a comprehensive blueprint for this new age migration in the ensuing chapter.

The Natural Distribution

In a small town in Germany, in the year 1777, a child was born who would one day change the world of science and mathematics. His name was Carl Friedrich Gauss and is considered as one of the most brilliant mathematicians and scientists of all time. Gauss was a child prodigy. He could do complex mathematical problems in his head, and at quite an early age, he was able predict the orbit of a comet before it was even discovered by astronomers. As an adult, Gauss made groundbreaking contributions to fields as diverse as mathematics, physics, and astronomy. However, Gauss' one of the most widely used, but still a relatively underrated, discovery was of a probabilistic distribution known as the normal distribution or Gaussian distribution. This distribution follows a bell-shaped curve as shown in Figure 8-1. It is used quite often to describe distributions of many phenomena, such as distribution of heights of people across any demographic, distribution of rise and fall of stock prices, weights of newborn babies, outcomes of tossing a coin multiple times, and the list goes on and on. In simple terms, Gaussian distribution has some very elegant properties. First and foremost, it is a symmetrical distribution with the axis of symmetry and the midpoint of the curve being at the peak. The peak at the center of distribution means that the number of samples having the average value are relatively higher than the number of samples having values that are farther away from the average. Also, as we go away from the center, the number of samples reduces on either side. This entire distribution is defined using only two parameters, mean (μ) and variance (σ^2) (or standard deviation (σ), which is square root of variance). Mean is the value of the distribution at the center and variance defines the rate at which the value drops as we move away from the center in either direction. In other words, it gives information about the contrast in the distribution. If the value of variance is 0, then we get a flat or uniform distribution where all samples have the same value. As the value of variance increases, the distribution becomes more peaked with sharp drops on both sides.

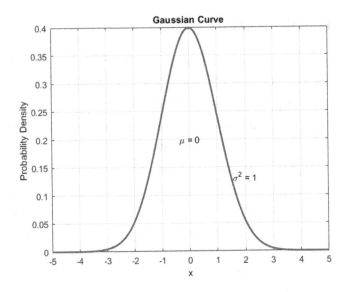

Figure 8-1. *Gaussian distribution*

Why is this distribution so special when there are hundreds of other distributions possible? The answer to this question lies in another discovery made by a rather lesser-known mathematician, Abraham de Moivre, which was further enhanced by Pierre-Simon Laplace. This discovery is known as "central limit theorem." In order to understand this concept, we are going to dive a bit deeper into mathematics, but bear with me, and very quickly we will come out to describe general concepts.

Central limit theorem provides a mathematical framework for a distribution of distributions. In technical terms, it states that if you take any distribution (not necessarily Gaussian, such as exponential distribution, Poisson distribution, Gamma distribution, etc.) and then take the average value of that distribution (also called mean) and look at the distribution of the average value across many distributions, the resulting distribution is always a Gaussian distribution as the number of distributions approaches a very large number. What this means is that no matter the type of process you are looking at, the parameters of the process will behave as per the bell curve when you have a large enough sample. This theorem makes a profound impact on statistical and probabilistic analysis and makes Gaussian distribution a de facto tool. Now, why would we be looking at Gaussian curve and central limit theorem in a book about human evolution and artificial intelligence? Let's come to that now.

Let's ignore all the theoretical statements and abstract mathematics and look at some real-life examples. If we gather a large set of data containing quantitative measurements of different features exhibited by humans, such as their height, weight, size of hands, fingers, even IQ, and so on. Consider a sample graph, as shown in Figure 8-2 that shows the distribution of height of all the humans on planet Earth. The X-axis or horizontal axis shows the height and the Y-axis, or the vertical axis shows the number of people having that height. If we plot a similar graph for all the other features, we end up getting the same bell-shaped curve with different values for mean and variance. If we expand our categories to include other biological processes such as blood pressure, enzyme activity, aging, or features of species that do not involve humans such as animal body length, leaf sizes on trees, microbial growth rates, seed germination times, and this list can go on and on, and plot their distribution, we always end up with Gaussian bell-shaped distribution. It is not just a coincidence, or some quirk shown by nature and mathematically we can prove using central limit theorem that it is the only way these features will distribute as long as they are not explicitly controlled. It's quite fascinating to see mathematics dancing in intimate harmony with nature.

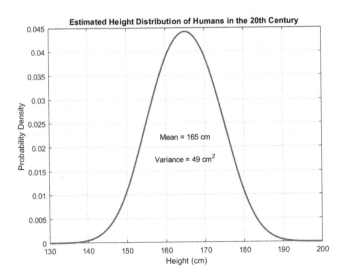

Figure 8-2. *The distribution is based on the results gathered by World Health Organization (WHO) and other studies*

As we move from these naturally occurring artifacts and look at man-made ones, things change quite drastically. One of the fundamental features that humans have created, and which is also at the center of the modern capitalistic world, is the amount of wealth possessed by an individual. One would estimate that wealth would also be distributed as per the Gaussian distribution as all the other physical and abstract human features such as IQ, height, weight, body strength follow the same distribution. A small chunk of wealth would reside with the richest and the poorest while the middle class led in the overall possession of the wealth. One thing to note here is that, when we are talking about the percentage of wealth, the number of individuals gets multiplied with the accumulated wealth, and as a result, there would be a bias toward the rich in the graph, because even if there are equal number of rich and poor people, the rich people's group will be multiplied with higher number. Figure 8-3 shows such distribution with slight skew toward the left or the riches.

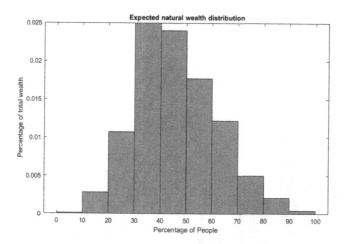

Figure 8-3. *Expected wealth distribution based on underlying normal distribution*

However, and it's a big however, the wealth possessed by humans does not follow Gaussian distribution! Not with some bias, not even close!! There are no exceptions to this observation across the world with all the variations with respect to race, religion, geographic location, technological advancement, and so on. Most commonly observed distribution for wealth is called as pareto distribution, as shown in Figure 8-4 (the plot only shows till the top 50%, but as you can see, the rest half of the population aggregately possess less than 4% of wealth that the columns representing their wealth would not be even visible). The crux of this distribution is that a large percentage of the wealth is concentrated with a very small set of individuals and the remaining large set of individuals have collectively a very small percentage of wealth, with the middle class just lying somewhere in the middle. Of course, each society would have some variation of this distribution, but nonetheless, the shape stays very similar in principle.

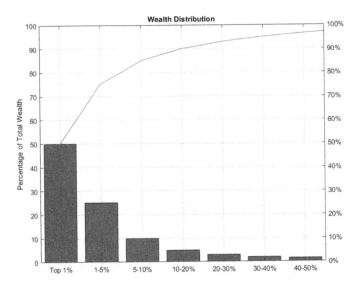

Figure 8-4. *Distribution of wealth in any society. The graph ends at 50%, as the columns for the remaining half of population possess less than 4% of the wealth and it would not be visible*

The population of any big country in the world is definitely a large number by any means and as such the central limit theorem should apply just as it applies to all the other features we saw earlier, and we should end up with normal distribution. But that is not the case! There is not even a hint of Gaussian-like distribution here. Isn't it an anomaly? Isn't it unnatural? Isn't it just plain wrong and a monumental failure of our economic system as a whole? When all the biological processes that drove the evolution of life forms on Earth leading to creation of humans follow a fundamental rule, a law of nature, humans have successfully broken it. This example just shows yet another outstanding achievement from humans. But is it for the betterment of humans? Is it a "good" thing to break this rule in this matter? The answer would be quite vehemently NO!! No rational person in their right mind can argue that it is "fair" to have one person living on streets with no food and shelter, while just a few hundred meters away another person is living in multiple million-dollar mansion, wasting plates full of food. Perhaps, the former person may have made some mistakes in their life, while the latter did a whole lot of great things; but are these differences justifiable enough for this extreme contrast? The more interesting question is whether these mistakes and good deeds are solely responsible for the differences in their current state, or have the man-made rules of our society played a role as well? On an aggregate level, human behavior would still follow the Gaussian distribution and as such the observed extreme contrast in wealth distribution must be attributed to man-made rules of the society and civilization, making them quite unnatural and unfair or unjust.

Origin of Money

Make no mistake, the current state of human civilization is still a huge improvement over what we had thousands of years ago. Starting with small communities of hunter-gatherers, who continuously fought against each other for resources, territories, we have certainly come a long way. At the hunter-gatherer stage, the primary objective for humans was to achieve the basic needs of survival, such as food, water, and shelter. The communities were continuously in search of these amenities. As they did not have effective ways to future-proof themselves for these amenities, the struggle was always ongoing as long as life persisted. Still, the efforts to improve on hunting and gathering and preserving the gathered material kept on going, and in time, they saw fruition in the form of numerous inventions as we discussed in the earlier chapters. Farming came into existence, thereby providing a steadier supply of food. Improved architecture and engineering technology provided more sustainable shelters as well as a steady supply of water over longer distances from rivers and lakes. With these inventions, we essentially came out of the hunter-gatherer phase. We were able to expand our civilizations in larger areas and provide all the basic needs in a more or less automated manner. However, thousands of years of persistence were etched into the human DNA and were not easy to get rid of. Even after having achieved the control of basic necessities, the hunger for more did not die. Instead of calming down, it found new opportunities in the form of insatiable greed for more. With the existence of sustainable basic resources, new forms of civilizations arose. The most powerful elites formed civilizations based on monarchy and took control of managing ever larger groups of humans. With more established civilizations, the barter systems from the days of hunting and gathering turned into currency-based exchange systems. With currency came the concept of wealth. Now instead of food and water, humans were hunting and gathering wealth. However, the mindset of hunting and gathering persisted without its original need and context. The ever-present fear for tomorrow, albeit unrealistic, kept on pushing the boundaries of how much to hunt and how much to gather. The greed for more kept on growing, as if it were avenging the age-old failures of prehistoric humans. With thousands of battles and wars fought at various scales ranging from small kingdoms to all the way modern world wars, the greed for more kept on going. However, since the second world war and its dramatic end with inexplicable destruction from nuclear bombs, humanity seems to have shaken up a little. As of the writing of this book, it's almost 80 years since the start of the second world war. Humanity is still quite far from being peaceful, but the situation is far better than it was in the last century.

We have certainly learned a lot from past mistakes, but we still carry some of the fundamental objectives that drive our day-to-today decisions from the era of hunter-gatherer generations. To put this into perspective, let's look at the times when it all began. The hunt used to be for food, water, and shelter. Gathering was primarily to future proof these needs so they can be used on the days the hunt is unsuccessful. However, for most part of this era, technology was not mature enough and most of these items were perishable and gathering was quite short-lived. It was quite common to have a barter system when there was an imbalance of these items across different people or even groups where goods were exchanged to satisfy the needs of both parties. The fairness of these trades would always be difficult to judge, but the system worked quite effectively for thousands of years. Fast forwarding to the times when civilization started to take root, the concept of money or wealth in general was born. It was created to form a standard reference in a complex trading environment. However, as one can buy anything with money and the more money you have the more you can buy, money became a single point of focus from the perspective of hunting and gathering. Furthermore, in ancient times, money was typically made of metal and had a very long life, meaning it was not perishable at all. Thus, we had invented an awesome tool as a culmination of our insatiable need to hunt and gather. Previously, due to the perishability of goods, there was no point in gathering them beyond a certain limit. But, with everlasting money made up of metals we would hope to gather so much as to suffice for one life or even the lives of future generations. This essentially removed all the brakes on our desire to hunt and gather. Gathering more money would not only give you the power to buy more goods, but also power to control other people who would be willing to do things for you in exchange for money. Thus, money stopped being just a reference unit for trading but also became a weapon to control people. It also crafted itself as a sole target for being stolen. It is better to steal money rather than goods, as goods can be perishable, but money would last forever. With money, you can buy whatever you want whenever you want. A question then arises as to how much money can there be? In the early times, money was tightly coupled with rare metals such as gold, silver, or gemstones. As any kingdom or country would have a limited amount of these "commodities," there was a limit to how much money there could be. Later, when it became too difficult to carry actual metal coins, paper money was invented; however, it was still backed by silver or gold, and the paper money was just a form of token that was approved for use instead by the country or kingdom. This type of money is called commodity-backed money. This definition of money clearly does not value services and other natural resources by default that one can provide or gather or tools or other artifacts that one can create and limits the scope of how much wealth that can be mustered by the civilization. Over time, as modern civilization started to take root, the limitations of commodity-backed money became apparent and the concept further evolved into what we use today. It is called a fiat currency. Although the oldest record of usage of fiat currency can be dated back to eleventh century China, the modern world saw its first use in the seventeenth century when Bank of Amsterdam issued paper money that was not backed by gold or silver. If the money is not backed by gold or silver or any other commodity, then the obvious question arises that who decides how much money there can be? To answer that question, let's look at the meaning of the word "fiat." It is a Latin word which means "Let it be done." This meaning is literally followed here. When a country creates a certain amount of money, it is declared as a legal tender by the governing body of that country. The value of the fiat money is based on the trust that people have in the government that issues it. First and foremost, fiat money lifts the limitation that money need not be dependent on the availability of fixed commodities and theoretically there can be unlimited supply of such money. However, with this flexibility, there come multiple risks such as the volatility in the value of money depending on the stability of the government, trust of the people on the government, or even the international trade situations. Typically, it is expected that the government would handle the money in the best possible manner, there are many examples where the currency of a country completely collapsed in a matter of months or even weeks, making all the money that was accrued by the people over their entire lifetime nearly worthless in a short period. Typically, the amount of fiat money that is created by any stable government depends on five factors: (1) Economic growth: When an economy is growing, there is a need for more money, and it is a reason sufficient enough for the government or central bank to print more money. (2) Inflation: Inflation is defined as a general increase in prices and as a result a decrease in the purchasing power of money. Inflation is another volatile entity and is rather unpredictable. However, when inflation is high, the government may create more fiat money in an attempt to stimulate the economy. (3) Government spending: The government may just decide to create more fiat money to finance its own spending, such as on social programs or infrastructure projects. (4) Financial or political crises: These are additional examples of volatile situations and can be quite different than inflation. In such situations as well, the government can create more fiat money in the hope of stabilizing the economy. Creation of additional money to fix a bad economic situation can itself become a root cause for making it worse creating a vicious circle.

Another implicit problem with any commodity-backed money is that the overall wealth is limited to availability of a fixed commodity. With such bounded wealth, all the citizens automatically become part of a zero-sum game. A zero-sum game is a situation where total gains and losses of the entire system add up to zero. In such a system, the gain of a person or entity becomes the loss of another person or entity, thereby pitting all the individuals against each other. In principle, it cultivates a competitive environment that makes it impossible to have a win-win situation. Not only does this create an inherent animosity, but it puts an unnatural limit on the amount of total wealth that can be created by the community. Any original creations or discoveries of natural resources do not really add to the measured wealth of society. Imagine two kingdoms with equal amounts of gold, but people in one civilization are extremely hardworking and they are producing high quality food and tools and shelter, while people in other civilizations are just wasting time. Should these two civilizations be equally wealthy? A commodity-backed system would say "Yes"! Even if this might appear obviously wrong to us, we are living in a very similar environment. Fiat currency tries to fix this by opening up the creation of additional wealth as needed. However, the creation of this wealth is managed by a single entity and as such the creation of more currency/wealth does not directly correlate with the new creations from the society, and in most cases, such additions of wealth only dilute the currency value, keeping the overall wealth unchanged in practice.

We have unknowingly bound ourselves into the golden bonds of shiny coins. With the creation of money, we did not really increase the potential of what humans can achieve, but we really limited it to a petty finite sum. The creation of money bereaved us of the infinite wealth that was provided to us by nature and by our own capabilities.

On top of the complications that came with fiat money, the financial industry saw yet another dimension of abstraction in the form of stock market or securities. A stock in essence is a small part of a company. A company can have thousands to millions of such stocks. People can buy and sell these stocks in a market called the stock market. Any company whose stocks are up for sale in a stock market is called a public company and it is now owned by all the people who possess its stock. It is a way to raise money for the operation of the company. The origin of stock market and overall concept of stock is highly debated, but some of the early examples include the founding of Dutch and British East India Companies in the early seventeenth century. The concept of stock essentially creates a layer on top of fiat currency that deals with the units or parts of a company. Now, instead of talking about the currency that is issued by the government, we are talking about the valuation of a company. Thus, depending on the performance of the company in the market, its valuation or stock price can fluctuate in terms of the underlying currency, which itself is also volatile as we saw earlier. Even though the East India Companies initiated the concept of stocks, the modern stock market was not founded until the nineteenth century when industrial revolution kickstarted the worldwide economic growth. First stock exchanges in the United States were formed at the start of the nineteenth century and specifically New York Stock Exchange (NYSE) was founded in 1817. Soon, managing money became synonymous with managing investments in the stock market. As mentioned earlier, the insatiable greed of hunting and gathering was feeding at an even greater rate with the stock market. After accumulating the money, if the money is kept in banks, it sees rather limited growth, but if you invest it in stocks, there is potential to increase your money at a much faster rate without you really doing any work. The companies and their employees are doing all the work, and if you can just somehow identify a company that is going to do better than other competing companies in near- or long-term future, you invest your money in their stock, and it grows with the growth of that company. However, in order to understand the market and to predict systematically which company is doing better, one needs to perform extensive research, which itself can cost a lot of money. Without such research, one can only try luck or make a little more educated guess based on partial information that is available. However, in comparison with organizations that are spending significantly more efforts in market research, the guesses based on pure luck or partial information are always going to lose in the long term and there is no need for any elaborate equations to prove that. Thus, only the rich can understand the stock market better and can get even richer. To make the system more complex, there are concepts such as mutual funds, and option-chains. A mutual fund comprises stocks from multiple companies at different proportions. One can now buy stock of a mutual fund essentially buying fractions stocks from hundreds of companies with a certain proportion. The options are another level of abstractions on top of individual stocks. The fairness of the economy has taken yet another huge hit with the evolution of the stock market. It should not come as a surprise, as the very origin of the concept of stock also lies in the foundation of the East India Companies that served primarily the elites.

At the start of the twenty-first century, a new type of currency became popular, known as cryptocurrency. Cryptocurrency is a digital or virtual currency that uses cryptography[1] for security. A defining feature of a cryptocurrency, and arguably it's the most inviting one, is its organic

---

[1] Cryptography is the practice and study of techniques for secure communication in the presence of third parties. It is the art of transforming original information into an alternative form (referred to as ciphertext) so that only those for whom it is intended can read it.

nature. It is not issued by any central authority or government, rendering it theoretically immune to political interference or manipulation. It can also serve as its weakness, as the only entity that can determine some of the founding parameters of the currency are created by individuals or the group of individuals who started the currency. In general, these individuals cannot be considered as trustworthy as any government of a nation. However, if a large number of people start supporting such currency, trust can grow. This being a purely digital and distributed currency, it is also not restricted to a particular nation or group of nations in principle. However, individual nations may restrict its use. Cryptocurrencies use decentralized control as opposed to centralized currency and central banking systems. The decentralized control of each cryptocurrency works through a blockchain, which is a public transaction database, functioning as a distributed ledger. This concept makes the currency extremely transparent where all the transactions are available to all the people using the currency, making it extremely difficult to fraud or counterfeit. The first cryptocurrency, Bitcoin, was created in 2009 by an anonymous person or group of people under the alias Satoshi Nakamoto, likely in Japan. Bitcoin was designed to be a peer-to-peer electronic cash system, meaning that it could be used to send and receive payments without the need for a third party, such as a bank. The value of cryptocurrency is based on a number of factors, including: (1) Scarcity: There is a limited supply of most cryptocurrencies, which means that they are scarce. This scarcity can drive up the price of cryptocurrencies. (2) Demand: The demand for cryptocurrencies is also a factor in their valuation. If there is a lot of demand for cryptocurrency, the price will go up. (3) Utility: The utility of a cryptocurrency is also a factor in its valuation. If cryptocurrency is useful and people start using it to buy and sell goods and services, more people will use it, which will drive up the price.

The creation of money in cryptocurrency is done through a process called mining. Mining is a process of solving complex mathematical problems in order to verify transactions and add new blocks to the blockchain. Miners are rewarded with cryptocurrency for their work. The trades of cryptocurrency are done through a process called trading. Trading is the buying and selling of cryptocurrency on exchanges. Exchanges are websites or platforms where people can buy and sell cryptocurrency. Cryptocurrency also comes with its own set of risks, including: (1) Volatility: The price of cryptocurrency is volatile, which means that it can fluctuate wildly. This can make it risky to invest in cryptocurrency. There is no governing body that can stop the transactions if the value starts altering too fast the way fiat currencies and stock markets are governed. Also, one can transact in cryptocurrency 24/7 unlike banks or stock markets that close at certain times. (2) Security: Cryptocurrency is a relatively new technology, and there are still security risks associated with it. For example, if your cryptocurrency wallet is hacked, you could lose all of your cryptocurrency. (3) Regulation: The regulation of cryptocurrency is still evolving. This means that there is some uncertainty about how cryptocurrency will be regulated in the future.

In principle, cryptocurrency tries to solve some of the fundamental flaws in regular "fiat" currency used by all the nations, but it has its own drawbacks. The most important one is the basis for creating more money. The fact that it is based on the work that is performed by machines makes it quite arbitrary. Another limitation of cryptocurrency is that the maximum number of its coins that can be mined is limited to some arbitrary number (in case of Bitcoin, it is twenty-one million[2]). This limit is hardcoded into its protocol and cannot be changed. It makes the currency scarce and valuable, but it also adds another arbitrary limit on the maximum allowed wealth. Many others followed the concept to create their own cryptocurrencies, but most of the popular ones do have a similar limit on maximum currency. These two drawbacks limit the scope of

---

[2] There are multiple arguments on the specific number for the limit on the number of bitcoins, one of them being: this number is a reference to the number of atoms in one ounce of gold.

cryptocurrency and likely it will never replace the fiat currencies that are issued by the governments.

In recent years, around the second decade of the twenty-first century, another new concept of ownership of wealth has emerged, called NFTs or non-fungible tokens. These are also fully digital assets that use similar block-chain technology as is used in cryptocurrency, but in this case, it is used to record ownership. The ownership can be of anything from digital art to real estate. NFTs are globally unique and cannot be replaced or duplicated ever. This makes them quite attractive from the perspective of being a collector. They do represent wealth, but there is no systematic mechanism for its creation and valuation. As such, it stays in its unique position alongside the other currencies.

Overall, in order to fix the problem with commodities-backed money, we have created a monster of a system that is quite volatile and more importantly extremely complex. In order to keep this currency system stable and balanced, it needs thousands of people and many organizations across the nation and world to continuously monitor its status. This led to the creation of banks and similar financial institutions. Then we expanded into a multinational multicurrency complex ecosystem. In order to stay on top of the volatility of the value of money, one must keep investing the money to keep its value updated with changing valuations. With all these complications, it created a whole new avenue for job profiles such as money managers that manage money for other people, financial advisers that educate people about managing their own money, financial analysts who analyze the volatility in the market, actuaries who are niche mathematicians that use mathematical and statistical techniques to come up with insurance premiums and so the list goes on. The abstraction from the original concept of money as a reference unit for trading became so complex that it became a thing in itself. We are currently living in such an abstract and contradictory space that, on the one hand, we understand the advantages of having more money, but when we don't quite achieve our target, we tend to tell ourselves that money is not everything and it's just a tool to buy the stuff that we really need. One would think that with such an advanced, elaborate and complex system of managing wealth, it would make it truly just and fair and almost automatic to use for most people. However, the result seems to be exactly the opposite. There are so many jobs in this field that are extremely high paying that mean nothing when it comes to satisfying the basic needs of people. Rather, most of the jobs that are directly responsible for satisfying the basic needs are some of the least paying jobs. All the gravy is getting sucked by the superficial jobs that are created by the overly complex system that has become more and more complex over the years. Instead of making the system better or fairer, the increasing complexity of the system has only helped the people who are involved in making it more complex and as such led to the creation of a vicious circle that is feeding on itself.

To take a concrete example, as per the report from Pew Research Center, the top 1% earning population possesses 32.3% of an entire nation's wealth in 2021, the same number was at 30.8% in 2020. The trend has been going on for decades and has accelerated in recent years. The primary reason for this extreme contrast is due to the overly complex financial system including the stock market giving unfair advantage to the rich. Furthermore, this accumulated wealth gets passed on from generation to generation, thereby giving no chance to the poor to catch up. In recent years, the worker unions have also declined, reducing their control over the wages. These findings are not secret, and the government tries to fix the issue by updating the tax structure to penalize the wealthy, increasing the minimum wage, etc. But the tax structure itself is also quite complex and it is a common practice for riches to find the loopholes in it and make the tax increases meaningless. The increased minimum wages are also typically in such a low bracket that they barely enable the workers to survive. These measures are extremely superficial and do not really address the core issues with our capitalistic economic system. There is always a limit to how far it can go. Once it crosses the threshold, it will be greeted with social unrest and civil wars. The time is soon going to come when we will have to make a fundamental change in the system, and I believe Artificial Intelligence can come to help.

Rise of Civilizations

Along with the economic principles, the rise of civilizations also led to the creation of sociopolitical principles and ideologies. Over the past thousands of years, we have made significant improvements in this aspect as well. We started with simple communities of hunter-gatherers where there were no well-defined roles or structured organizations. With the establishment of farming and settlements came a more coordinated approach toward the communities and soon a primitive social ideology was born in the form of chiefdoms. Chiefdom is one of the most primitive sociopolitical systems in which there is a single chief who rules over the community. Chiefs are usually chosen by the people, and they have the power to make decisions for the community as well as enforcing rules or laws of the community. As the communities grew, chiefdoms evolved into kingdoms and empires. In a kingdom or an empire, there would be a single leader, also called a king or emperor, similar to the chief in the chiefdom, but with more power and a bigger community to rule over. Typically, the king or emperor would rule the entire society that is part of the kingdom or empire along with a group of select elites and the leadership would typically pass from generation to generation by the way of naïve heredity. This system is also called Monarchy and is not really considered as a sociopolitical ideology by modern definition, but it was a steppingstone that led to the more elaborate and thoughtful sociopolitical systems. Typically, it is considered that such systems existed till about the ninth century in Europe, coinciding with the fall of the Roman empire. However, similar systems still existed in pockets across the world for many centuries after. For example, India had many small kingdoms till the eighteenth century when British invasion gathered all the Indian subcontinent under the same leadership, or China had its historical monarchy of Qing dynasty as late as the twentieth century. Around the fall of kingdoms came a system called Feudalism in Europe which lasted till around the fifteenth century, that paved the way for more modern sociopolitical ideologies such as capitalism.

With each new system, people tried to fix the issues and drawbacks of the earlier ideologies; in many cases, the new systems did address them but created new issues of their own, or the solutions offered did not pass the test of time. Nevertheless, these ideologies provided a framework for organizing society, determining power structures, and defining the relationship between an individual and society. We see quite a wide spectrum of such ideologies in the history of mankind, which have ultimately converged into a smaller set. Even in present times, in the early twenty-first century, we have at least a couple of examples of different ideologies in action across different nations.

Capitalism

On one side, also called the *right* side, we have capitalism. Capitalism is essentially a market-based economy that promotes private ownership and assumes the automatic distribution of the resources based on supply–demand balance in the market. Although most modern capitalistic systems do start with the declaration of equality, equality is restricted to having equal basic rights such as liberty and the pursuit of happiness, etc., but for everything else, things are not defined and left for the system to automatically manage. The ownership rights in capitalism extend to everything that is required to carry out the business operations such as land, capital, other natural and man-made resources. The notion of automatic distribution avoids the need to have any governing body to decide who gets what. With such a fairly open-ended setup, the ultimate motive for any individual boils down to profit. This encourages all individuals to be competitive and results in motivating them to do their best. In theory, this should lead to innovation, efficiency, and growth. A governing body is still needed to avoid abuse of the system, but with minimal supervision. The role of the government is limited to enforcing the rights of all the individuals, ensuring fairness of the trades and contracts and maintaining healthy competition by providing a legal framework. Capitalism gives substantial freedom and a wide range of choices to all the individuals to pursue their interests, choose the occupation they want as well as how to spend the wealth they have earned; all within the confines of the legal system. It almost looks like a perfect system. However, with all these benefits come some drawbacks as well. One of the most fundamental drawbacks of capitalism is the creation of classes. It is important to remember that capitalism does not define classes anywhere in its definition, rather it focuses on equality. However, as we have seen in all its implementations in Europe as well as in the United States of America, class structure is an unavoidable outcome of this system. Overall, in the early years, capitalism showed a strong promise as all its benefits appear to work and bear fruit, the distribution of wealth in the early years of capitalism is not as extreme and leans more toward natural distribution as we discussed earlier, essentially, the majority of the people in the society are happy. However, even with the government moderation, the system starts to become more and more unfair and unjust and leads to an increased contrast between the classes leading to pareto type wealth distribution. Money starts to get concentrated into smaller and smaller groups increasing the inequality, and the trend is strongly in one direction with no natural way of reversal. Ultimately, it can lead to social unrest and civil wars.

Communism

On the other side, typically called the *left* side, we have communism. Russia and China represent the prime examples of the countries following this doctrine other than some smaller implementations in pockets across the world. The present-day communism has evolved over a couple of hundred years since it was first proposed by German philosophers Karl Marx and Friedrich Engels in the mid-nineteenth century. Though originated from the same source, there are some variations in its implementations as we see in Russia and China. Communism is the most modern system based on the chronological evolution of systems and as a matter of fact, Karl Marx specifically proposed it to overcome the drawbacks of capitalism. The primary driving principle of communism is the creation of a classless community. Marxist communism specifically identifies two classes of society as bourgeoisie or owners of production and proletariat or the working class. As per Marxist communism, the means of production are owned collectively. Wealth and resources are distributed based on the principle "from each according to his ability to each according to his needs." Communism assumes that all the individuals of society are equal and therefore they all should be treated equal. There is no scope for hierarchy or classes. It also strongly advocates the abolition of private ownership of natural as well as man-made commodities and promotes the idea of collective ownership. Equality extends to all aspects of life such as access to resources like food, shelter, healthcare, education. Even the opportunities one can get are forced to be equal regardless of the social or economic background. In order to achieve this, it suggests the use of central planning where all the key decisions about the production, distribution, and use of the resources are made collectively. Communism also promotes the concept of the state. The concept of state here is quite nuanced and rather complex. It is typically referred to as the dictatorship of the proletariat or the entire working class. In this context, the state is an instrument of class domination and is utilized to suppress the bourgeoisie or capitalist class and defend the working class. The definition of the concept of state may seem contradictory to its own principle of equality, but communism uses the state only as a transitional mechanism to suppress existing capitalist class till the entire community is converted into a single working class, the ultimate realization of the communist principles. It is a grand expectation of Marxist communism that once the transition is complete, the state should automatically dissolve. Till then, the state has full authority over all the decisions and is controlled by the ruling party. In reality, none of the communist countries have been successful in the dissolution of state. Rather the state provides the ruling party a backdoor toward dictatorship under the pretense of equality, giving them uncontrolled power. As per the wise words from British historian Lord Acton, "power corrupts and absolute power corrupts absolutely," the state has never been able to let go of the power. In spite of its daunting drawbacks, communism proposed some really novel concepts, that were ahead of its time. One of them was the true internationalism. The communist manifesto ends with "Workers of the world, unite!"

There are many other ideologies whose principles lie somewhere in between these two left and right extremes. Many aspects of all these ideologies seem fair and just on their face value, but the critical component that is missing in these definitions is their implementation and the people involved in governing these systems. Communism enforces equality to the extreme. It is true that biologically all the new-born babies are 99.9% identical as per the DNA variations, but since then, each baby goes through different environments and experiences, and as such their brains are quite different by the time they are adults. Also, pursuing the individuality of uniqueness is one of the key aspects of human life that separates us from the lesser species. Communism tries to enforce a uniform distribution of everything in society from the amount of food one can eat, the amount and type of work one must do, and the amount of wealth one must have. A uniform distribution is also theoretically a Gaussian distribution, but with zero variance. As we saw earlier, humans by nature are not born with uniform distribution, and forcing them to be equal in all aspects is not natural. Also, a desire to be different from others gives a purpose and motivation to human life. Communism tries to kill this at its root.

Intellectual Humanism

Combining capitalism and communism and the lessons that we have learned from the implementations of each, together with the omnipresent SuperAI, we are ready to march toward the next generation of the sociopolitical system. As we aim for thousands of years into the future with the new age civilization, we need to stretch the bowstring of imagination thousands of years in the past to ensure our aim has enough strength to reach the target accurately and authoritatively.

Capitalism led us on the path toward innovation and, over the last many centuries, we have seen the incredible impact of the innovation across the world. However, it came with division of society into classes and the valley between them kept growing deeper and deeper over time. Capitalism also made us more self-centric rather than community-centric. It is the underlying assumption behind capitalism that if all the individuals in the society succeed then the community must succeed. However, the success was not distributed naturally (think bell shaped), but it was skewed toward the upper class and over generations the disparity kept growing. Hence, even if we saw huge progress in innovation and technology, as a society the picture is a mixed bag. The problems that we see today in the form of non-sustainable use of natural resources such as underground water, oil, and gas; even disproportionate levels of fishing and hunting; that created deep and, in some cases, irreversible impact on the environment such as increase in carbon dioxide levels in the atmosphere, endangering the oceanic ecosystems, thereby endangering lives of millions of people. The sole reason for this situation is the total disregard toward the negative impact on environment and in turn humanity for advancing the profits of the select few industries and individuals. The legal system and governing bodies that were supposed to control the business activities in capitalistic societies failed miserably.

Communism tried to address this issue by putting the society back at the center, but in doing so, it enforced unnatural levels of equality and disregard for individual growth and happiness. On top of that, the concept of state, the entity that would facilitate the transition from capitalistic and divided society into the uniform worker class society, remained vague and created a backdoor toward dictatorship.

Combining the virtues of these systems we need to build a new system that keeps the growth and prosperity of the entire humanity at its center and along the way preserving individuality, individual growth, and happiness by providing appropriate priority to it. Here are the fundamental principles on which we can build the new age of civilization, let's call it intellectual humanism. Intelligence and humanity are both founding pillars of this system. The notion of intelligence here is not the superficial notion of individual IQ, but it is the collective wisdom of the human species that made it superior along with the intelligence of the machines that we have developed. We will couple that with the age-old principles of humanism that are etched on our DNA, that unite all of us on the global Earth to create a most natural sociopolitical system that should pass the test of time.

Here are the fundamental principles on which it would be based:

1. *Humans first*: Civilization based on intellectual humanism would be primarily rooted in the philosophy of humans first. All humans across the Earth and beyond (if humans have already diversified on other planets at the time) would be at the center of it. There are many causes of human suffering on Earth, but the top factors at the root of it are poverty, war and conflict, disease, discrimination, and lack of education. We will address each one of this through the fundamental principles of intellectual humanism and the specifics of the organization of the global civilization that would be explained later in the section.

With the help from omnipresent SuperAI managed by humans as detailed in the earlier chapter, all humans would have a guaranteed supply of basic needs as their birthright. This would include food, water, shelter, and basic medications as needed. "Humans first" philosophy does not stop there; but all the uber decisions that are made would have an aspect of how it helps mankind as a whole as the primary guiding parameter. These decisions include use of natural resources only in a responsible manner that replenishes them automatically as well as interaction with other species such that there is always a form of symbiotic relationship.

2. *Equality*: Starting from birth, all human babies would be treated equally, they all will have access to similar levels of resources in upbringing, education, sports, and all the other curricular and extra-curricular activities that are available at the time. Throughout their education, keen attention would be paid to their individuality and their innate preferences, natural inclinations and personal goals. Each year they would be given optional areas to choose based on these factors to help them achieve the best and satisfying education they can. The ultimate goal of the education is to prepare them to be a proud citizen of global intellectual humanity and to contribute to its success in the best possible way they can.

3. *Transparency*: Monitored by the omnipresent SuperAI, there would be complete transparency in all the operations carried out by humans as well as machines. Privacy would still be respected during private hours and inside the homes, but all the public areas would be under continuous surveillance powered by SuperAI ensuring continued law abiding and crime free environment. Transparency at work would ensure continuously fair treatment of the workforce and their corresponding rewards.

4. *Unlimited opportunity*: Starting with education, all the citizens would get regular updates on their progress and suggestions for their next steps. However, each individual would have an option to choose what they want to learn or work on. With help from SuperAI, each individual would be presented with a comprehensive set of options for the work that is available. The availability of the work can be bucketed into three parts: (1) Work that is necessary for the continued operation of the system. (2) Work that is explicitly requested from other members of the civilization in the form of entertainment, artforms, and so on or (3) work that is entirely novel and is purely the imagination of the individual (with proven positive impact on the civilization). There would be some minimum level of mandatory work that would be expected out of each individual based on their capacity and interest. Based on the help provided by SuperAI, it would be quite minimal. After that, each individual would be free to help civilization by contributing in their own ways.

5. *Fusion of classes*: By definition, all individuals in intellectual humanism would be an individual contributor or worker as per the Marxist definition, but at the same time, they would also be the owner of the work they are performing. All the organizations would have a flat structure with no explicit hierarchy. Seniority would be respected and would reflect in their rewards, but management of the workforce would be entirely handled by SuperAI with its omnipresent surveillance and coupled with peer feedback. Such a setup would encourage each individual to really focus on what they really enjoy doing. Any person can change their line of work at designated intervals spread at reasonable time intervals ensuring that continuity of work is maintained. Entrepreneurship or starting a new venture would be highly encouraged and all such requests would be evaluated at regular intervals. Based on peer reviews and interested followers, a new venture can begin and grow. Such a setup would truly create the single class culture that should feel natural and liberating for all the individuals.

6. *Distributed economy*: The notion of fiat currency or commodity-backed currency or even cryptocurrency would have to sunset in the light of the new distributed credit system. This system would exploit the learnings from the implementations of these systems but would improve on them. All individuals would get a fixed set of credits at birth, and they would replenish at a regular rate throughout their

early childhood and education. They would be expected to contribute back to civilization by doing carefully designed age-appropriate mandatory duties (again with possible options to choose from) during their education. Once they complete their education, they would be given a choice from three buckets for their career as described earlier. The range of rewards would be carefully drafted to ensure fairness, at the same time matching the efforts and impact of the work. The rewards would be in the form of credits that would be directly associated with the individual and would be tracked by SuperAI using biometrics without any need for explicit banking operations. The credits can be spent similar to the way we typically spend money for purchase of goods or services. The handling and tracking of credits would be entirely automatic and no additional effort need to be spent on it. Another important aspect of this economic system is the absence of inheritance. All the unused credits earned by an individual during their lifetime would be recaptured by the system. In intellectual humanism, each human baby would start from scratch as they enter into the world and would carve out their own identity just like any other baby. This is a crucial aspect of the economic system that is required to balance the distribution of wealth in a natural way. Imagine a world where all humans are working for the betterment of their own self or for others in the way we can and getting rewarded appropriately; just by doing things they are naturally born to do without worrying about competition. The greater the population, the more wealth is generated, and more people are helping each other making the civilization stronger and richer and happy. The new concept of wealth would eliminate the unhealthy competition at its root to make for road to true progress.

Equipped with these principles, we can start painting nuanced details of the new age civilization. We need to start by defining that the entire humanity would represent a single nation as per today's standards. We need to break the boundaries of the national borders and let the entire humanity embrace each other without having to go through the hurdles of Passports and Visas and so on. We still need to ensure that there is global accountability for actions of all the humans and it would be handled by biometric identification of all the humans through the omnipresent SuperAI in an entirely automatic and non-invasive manner. The world would be a giant conglomeration of self-sufficient colonies (SSCs) spread across the world. The SSCs would resemble the concept of a state in most countries such as the United States of America, India, etc. at the current time. Each SSC would be self-sufficient from the perspective of basic necessities of all its inhabitants but would still be an integral part and an irreplaceable cog in the global system of civilization. The SSCs would be connected via inter-SSC business operations just as there are inter-state operations.

Each SSC would have its own set of governing bodies operating on the same flat structure as outlined earlier. All the representatives would be elected by the members of the SSC. All the governing body meetings and operations would be accessible publicly to establish full transparency. The decision-making process would be aided by SuperAI taking into account votes from each member. Each SSC may have their own set of laws as an extension of the global laws based on the local specificities. SuperAI can ensure that all the laws originating from global scope to SSC are coherent and there are no discrepancies. Travel across the SSCs can be carried out using personal credits for pleasure or based on optional work transfers.

Each individual would also get an option to migrate to another SSC after a certain period. These migrations would be handled by SuperAI by optimizing the resources on a global scale.

The all-encompassing intellectual humanity-based system cannot appear suddenly out of the blue. It needs to start with the establishment of the first SSC, and then the success of the first SSC would drive many other states and countries to join one after another. Till the entire world is not part of the new system the intellectual humanity-based conglomerate would act as one nation and deal with other nations appropriately.

Once the transition is complete, the world will truly enter the new age of human civilization, appropriate for the time and deserving for humans, who mark the epitome of evolution by natural selection. We bow to Nature and respectfully take charge of the next steps. Evolution has served its purpose and has created the most advanced species, not just on the face of Earth, but on the face of the entire universe and it must rest now in peace as humans, empowered with their intellect and SuperAI take the charge!!

Conclusion

This chapter takes a radical turn in a different direction compared to all the other chapters in this book. However, it builds on everything that is presented earlier. It presents a possible future for humans that is near ideal, and it is only possible with the help of SuperAI. All the philosophers, theorists, and leaders in history probably dreamt of humans living like this, but could not realize it due to lack of an entity that is intelligent but does not have its own ulterior motives. Distributed control over SuperAI in a democratic manner can solve all the fundamental problems in creating a dream human civilization in the form of intellectual humanism and provide a path for humans to be true universal citizens!

CHAPTER 9

Final Words

Artificial intelligence is a fascinating topic, but it's the tip of an iceberg that has a lot hidden that does not meet the eye. In this book, we tried to go deeper to understand the concept from its roots. As we start diving deeper, the scope of the topic keeps expanding and as we saw in the last couple of chapters, AI is going to redefine human life in ways that history has never seen.

Humanity has come a long way, but from the look at the future, we are just getting started. The journey forward is a blank canvas, and it is up to us how we want to paint it. We can keep focusing on individual growth the way we are doing now under the pretense of capitalistic sociopolitical realms. However, it is not the natural way and is creating a divide in society that is just going to get worse till it reaches a point where it breaks. At that point there would be worldwide social unrest followed by civil wars with far-reaching consequences. This would take humanity hundreds of years back in time with the loss of progress made in all those years. The new system that would replace the existing one may not be an improvement. This is because it would not be created with long-term progress in mind. Instead, it would be driven by a motive of vengeance, a pattern we have seen in history one too many times. When there is hunger and poverty and depression and misery, all the focus is toward addressing these basic needs. The human mind does not function at its best in such situations. All the historical times that showed great progress in science and technology

© Ameet Joshi 2024
A. Joshi, *Artificial Intelligence and Human Evolution*,
https://doi.org/10.1007/978-1-4842-9807-7_9

were the times of prosperity, the times of happiness and satisfaction. It is in these times when we should not be complacent but show ardent fervor toward dispassionately criticizing the norms and making improvements. It is in these times when best decisions are made and implemented. The worst of capitalism is still some time away, and we should not wait for it.

Intellectual humanism is a thought exercise, a dream, a fantasy about the future from a positive mindset. It is an iconoclastic proclamation toward the betterment of humanity with deeply rooted faith in science and technology that is built on the sum of the facts from known history. It may not be the only option, but it is certainly an option. I would like to conclude this book by sowing this seed in the minds of all the readers.

To be continued...

Index

A

Abacus, 148
Advanced navigation tools, 86
Aeolipile, 93, 94
Air-based transportation, 191
Algorithms, 131
Alien life, 17
Alternating current (AC)
        electricity, 98
Amazon forests, 40, 41
Ampere's law, 99
AND operation, 153
Archimedes' screw, 85, 86
Artificial intelligence (AI), 1, 239
    capabilities, 2
    consumer needs, 4
    movies, 1
    predictions/theories, 3
    scenarios, 1
    technical knowledge, 2
Artificial neural network
        (ANN), 135
Astrolabe, 86–89
Atomic clocks, 90
Australopithecus afarensis, 72
Automatic distribution, 227

B

Big Bang, 7
Biological processes, 15, 211
Bitcoin, 222, 223
Blockchain, 222, 223

C

Capitalism, 227, 228
Capitalistic economic system, 225
Carbon, 149
Carbon dating, 44
Carell's experiment, 25
Celestial events, 87
Central limit theorem, 210, 214
ChaptGPT, 142
ChatGPT, 2, 4, 128, 143, 157,
        158, 163
Chiefdom, 226
Chinese chariot, 117
Circular argument, 15
Cisco fish, 65, 66
Civil engineering, 179
Civilizations
    capitalism, 227
    chiefdom, 226

© Ameet Joshi 2024
A. Joshi, *Artificial Intelligence and Human Evolution*,
https://doi.org/10.1007/978-1-4842-9807-7

U

V

W, X, Y

Z

Printed in the United States
by Baker & Taylor Publisher Services